Call of the Vine

Exploring Ten Famous Vineyards of Napa and Sonoma

by

Dr. Liz Thach, MW

Foreword by Tim Mondavi

MIRANDA
PRESS

Call of the Vine

Exploring Ten Famous Vineyards
of Napa and Sonoma

Text Copyright © 2014
By Dr. Liz Thach, MW
Published by Miranda Press
An Imprint of Cognizant Communication Corporation
18 Peekskill Hollow Rd P.O. Box 37, Putnam Valley, NY 10579-0037 USA
www.cognizantcommunication.com

This is a work of non-fiction. Names, characters, incidents, and dialogue are based on actual interviews with the participants as well as document analysis. The publisher and the publisher's agents represent that the data provided were formulated with a reasonable standard of care. Except for this representation, the publisher makes no representative or warranties, express or implied.

Call of the Vine: Exploring Ten Famous Vineyards of Napa and Sonoma / Liz Thach
p. cm.

ISBN: 978-0-9715870-5-2
BISAC: Business & Economics / Industries / Agribusiness

Printed in the United States of America

Praise for Call of the Vine

"A unique look at viticulture practices in a story-telling format. The descriptions of the vineyards and in-depth conversations with vineyard managers and winemakers create an entertaining read, and should be enjoyable for both wine students and enthusiasts."

> - *Dr. Linda Bisson, Maynard A. Amerine Endowed Chair in Viticulture and Enology, UC-Davis*

"A down to earth book that describes the joys and challenges of grape growing in two of the most magnificent regions in the world – Sonoma and Napa."

> - *Rich Thomas, Emeritus Professor of Viticulture, Santa Rosa Junior College*

"This book includes many useful details about viticulture and winemaking, and is presented in a first-person format that is engaging and easy to read. I enjoyed learning about the history of these great vineyards, as well as the interplay between nature, science and practical common sense that is so clearly evident in the daily lives of the vineyard workers."

> -*Dr. Mark Greenspan, President, Advanced Viticulture Inc*

"This book will appeal to wine students, vineyard and winery owners, and wine enthusiasts who are interested in learning about the nitty gritty details of vineyards and what makes a great vineyard in California."

> – *Elliott Mackay, CEO, The Wine Appreciation Guild*

Table of Contents

"Nothing else was left, not a coin. Nothing except...the bond with the vine upon him."

Idwal Jones, The Vineyard

Foreword by Tim Mondavi

My father always said, "Fine wine has an affinity for family." It was my grandfather's sense of discovery that brought our family to wine and to Napa Valley, starting in 1919. My father added his vision and zeal to elevate this region to the world stage. Today, at Continuum we build on this legacy to create a world-class estate, aided by the clarity and power of focus. Over time, I've discovered that the fine-wine business is measured in generations rather than in quarters. Therefore, I am very pleased to write this forward for *Call of the Vine: Exploring Ten Famous Vineyards of Napa and Sonoma*. The book celebrates for me all that is important about respecting the generations of the past and future through tending the land and its vineyards.

I have had the good fortune to grow up in the Napa Valley and have watched it grow into a world-class wine region. When my family opened Robert Mondavi Winery in 1966, it was the first new winery to open in the valley since the repeal of Prohibition. My professional career spans forty years, during which I have worked with the best vineyards in California. Additionally, I have had the good fortune to work alongside some of the greatest winemakers in the world, and as a result I realized that fine wine is the result of a great site diligently cared for by a passionate team—a harmonious interaction between man and nature. The Frescobaldis in Tuscany, growing wine for more than eight hundred years, demonstrate what can happen when you blend great vineyards and a family with a long-term view. Baron Philippe de Rothschild in Bordeaux proved that wine truly is an art in a way previously unheard of, compromising nothing in his pursuit of excellence.

Closer to home, I have seen remarkable winegrowing right here in California. In Sonoma, Hanzell Vineyards, planted in 1953 by visionary James D. Zellerbach, has the oldest Pinot Noir vines in the United States.

Nearby, Monte Rosso Vineyard, with soils quite similar to Continuum's, grows remarkable Zinfandel planted in 1890, high above Sonoma Valley. In Napa, there are fabulous vineyards, both mountain and benchlands, which showcase site-specific Cabernet Sauvignon in a variety of locations, such as Diamond Creek on Diamond Mountain, Stagecoach Vineyard on Atlas Peak, and my family's former vineyard, To Kalon in Oakville. Each of these has a unique sense of place, which, at its best, translates into a captivating and significant wine. For a great wine can only be made from a great site, with artistic commitment, to stand the test of time.

Our goal at Continuum is to produce a singular estate wine that merits recognition among the finest in the world. It is our aspiration to build solidly on my family's legacy and the unique qualities of this gorgeous site on Pritchard Hill. We don't expect this to happen overnight or in a decade or even in a single generation, but we are determined to achieve it, vine by vine and vintage by vintage.

Each of the vineyards profiled in this book has what is required to be a great site and a great growth. In the old world, the cru system reflects the belief that certain vineyards have the perfect combination of soil, sun, water, elevation, and aspect to grow fine wine. Kudos to Professor Thach for helping us to see the vinous riches that surround us here in California! It is my hope that as you read this book, you will come to understand all of the hard work that goes into tending a vineyard and gain an appreciation for how much heart and soul is shared by those who choose to nurture the vine.

Tim Mondavi
Continuum Estate
Napa Valley

Acknowledgements

There are so many people I want to thank and recognize for helping to bring this book to harvest. First is Mr. Gary Heck, president and owner of Korbel Champagne Cellars in Sonoma County. It was due to Gary's generous establishment of the *F. Korbel & Bros. Professorship in Wine Business*, an $18,000 annual research scholarship for three years, that I was allowed the extra time to research and write this book.

Next I would like to thank my publisher, Bob Miranda, who was not only kind enough to take a chance on an unknown wine business writer to publish my first edited book, *Wine: A Global Business*, but who also offered to publish this one. His steadfast support and encouragement throughout the years have been major blessings in my life. Likewise, his talented book designer, Lynn Carano, deserves much credit and appreciation for her brilliant design of the front and back cover of the book. Also to Michelle Drewien, founder of Zango Creative, and designer of the AVA and vineyard maps.

The visionary leadership of Mr. Tim Mondavi and his publicist, Burke Owens, are also greatly appreciated. Tim's viewpoints on the importance of place and the essence of land and terroir in creating a great vineyard establish an insightful and motivational overture for the book.

Much appreciation to all of the vineyard managers, winemakers, and public relations professionals who provided their time, wisdom and editing in making this book a reality. Without their support none of the stories of the vineyard could have been told. Likewise, I would like to express gratitude to the *Napa Valley Vintners*, the *Sonoma County Vintners* and the *Sonoma County Winegrowers* for providing the initial list of famous vineyards, and assisting with contact and historical information.

In addition, I would like to thank all of my students at *Sonoma State University* and other faculty and staff members who encouraged and supported me over the three years it took to complete this book. Special thanks to the *Institute of Masters of Wine* in London that initiated my deeper quest into the life of vineyards with their exam requirements to learn about the nitty gritty details of viticulture, and more importantly, why and how decisions are made to use specific clones, rootstock, farming techniques and the myriad of other issues explored in this book.

I would especially like to thank my family and friends for their support and patience with me as the manuscript evolved, especially my husband and daughter. Thank you also to my neighbors, Peter and Susan Goyton, for their feedback and editing. Special appreciation goes to my father, mother and grandmother Alma who have always taught me about wildlife, plants, and the importance of respecting and preserving the land.

Distinct recognition goes to all of those at the *University of Iowa's Writer's Workshop* who gave me feedback on how to revise the book, and encouraged me to write in the first person in a conversational format rather than the dry technical prose of the original manuscript, which ended up in the garbage can of a hotel room in Iowa. I am hopeful that readers will appreciate the more approachable style, and forgive me if I talk too much in the book.

Other muses who assisted me were my long-time vineyard mentor, Mr. Paul Dolan, who is quoted in this book, and two of my favorite writers. The first is Willa Cather, who has always inspired me with her stories of pioneers who settled America and her love of the land. She was one of the first American authors to describe the land as a main character. The second author is Kermit Lynch, whose book *Adventures on the Wine Route* was an inspiring model of how to write in the first person about wine. I savored every chapter of his book, taking the time to read it slowly so I could stretch out the pleasure. Though I don't presume that my book can be as pleasurable, I am hopeful that readers will take a pause between chapters so that the uniqueness of each vineyard can bloom in their thoughts.

My goal in writing this book was to share my love of vineyards and my home state of California, as well as to provide the type of technical vineyard details that students of wine and wine business need in pursuing degrees and certifications, or wine enthusiasts will enjoy as part of their ongoing learning. In the pursuit of this goal, I hope that the hard work, dedication, and love of the land expressed by the vineyard managers, winemakers, and vineyard workers in this book, can be appreciated and celebrated by all who read it.

Warm regards,
Dr. Liz Thach, MW
Sonoma County, California

Chapter One

The Soul of the Vineyard

It all started in France on a slightly cloudy morning in April. I was standing in front of the Romanée-Conti Vineyard in Burgundy, gazing at the famous cross that heralds this sacred plot of land, when a carload of Japanese tourists arrived. They quickly climbed out and arranged themselves in front of the cross and next to the short rock wall with Romanée-Conti engraved upon a slab of stone. Then they took turns taking photographs of one another, all the while chattering excitedly in Japanese and smiling widely into the camera as they gestured to the cross and sign. I stood back and watched until a slender girl approached me.

"Please, take photo of us?" She timidly held out her camera and motioned to the group of four people behind her.

"Of course," I said, reaching for the camera. Obediently, I snapped several shots of the beaming group.

"Thank you," the girl said as she approached me to retrieve her camera. "Very famous vineyard. Makes very expensive wine."

"Yes." I nodded and handed her the camera.

"Thank you." She bowed her head twice in a Japanese gesture of respect and rejoined her companions.

As I watched their car drive slowly away down the narrow road into the tiny village of Vosne-Romanée, I saw two more cars approach, and I watched in amazement as the occupants climbed out and took similar photos of the vineyard. Later, at the La Tache Vineyard down the road, other tourists arrived to take photos near the stone engraving on the rock wall heralding La Tache.

It was then that it struck me. As a fifth generation Californian working in the wine industry since 2000, I know there are many magnificent vineyards in California. Some are known for their historic roots going back to the eighteen hundreds. Others are famous for unique attributes or for consistently producing top award-winning wines, but these famous West Coast vineyards do not often receive the recognition they deserve. I don't recall seeing carloads of tourists stop to pose and take photos next to a California vineyard sign. But then again, it is not a common custom to build rock walls around California vineyards and post stone monuments proclaiming the name of the vineyard. Still, there are many famous vineyards in California, and they, too, deserve some respect and recognition.

It was then that the seeds for this book sprouted inside me, and I knew that eventually I wanted to write about the special vineyards of California. I decided to focus on famous vineyards of Napa and Sonoma counties because this is where I live. Furthermore, I have a small hobby vineyard of Pinot Noir and Sauvignon Blanc located in the Sonoma Coast AVA (American Viticulture Area). This area is blanketed by the cooling fog blowing through the Petaluma Gap—a gap in the coastal mountain range that allows the wind and fog from the Pacific to flow across vineyards in this area. The lessons I have learned from my own small vineyard are also a motivation to explore the great vineyards of my homeland.

Therefore, the purpose of this book is to identify and celebrate ten famous vineyards of Napa and Sonoma. This includes conversations with vineyard managers and winemakers who tend these sacred plots of soil, which produce exceptional and historic wines. These experts not only provide information about the unique soil, climate, elevation, and viticulture practices but also describe the history, story, and soul of the vineyard.

According to one ardent viticulturist I know, a vineyard can be referred to as a "community of living beings with a unified purpose."

Tapping into this purpose with a positive intention to help the vineyard achieve its best can be considered the highest goal of the vineyard workers and winemaker. Indeed, there is an old adage that says "the best wines are grown in the vineyard," and top winemakers "pitch a tent in the vineyard" to attune themselves to the vine. With this in mind, for this book, I ask each vineyard manager to describe the meaning of the vineyard, and I also attempt to do this myself.

Next, I spend some time with the winemaker to focus on a specific wine for which the vineyard is well known. We discuss winemaking techniques and how the vineyard contributes to the final product. There is another old saying that "Good wine tastes like fruit, but great wine tastes like a place." Through this lens, we explore how the "place of the vineyard" impacts the wine.

Finally, the resulting wisdom gleaned from these interviews is synthesized in the last chapter, entitled "Lessons from the Vineyard," which explores common themes and differences between the ten vineyards described in this book. This information is not only useful in determining what components help to create a great vineyard but is instructional to all who heed the "call of the vine" and are driven to plant a vineyard—like me.

Identifying Ten Famous Vineyards in Napa and Sonoma

With hundreds of vineyards and wineries in both Napa and Sonoma counties, it is no small feat to narrow it down to ten of the most famous. Therefore, I decided to seek some assistance by contacting Napa Valley Vintners, Sonoma County Vintners and Sonoma County Winegrowers. I sent the organizations an email describing the purpose of the book and requesting a list of three to five vineyards that fit the following criteria: historic vineyard; unique feature, such as location or varietal wine; and/or consistent production of a high number of award-winning wines.

As would be expected, the organizations responded that it was difficult to narrow it down to such a small number, but they each provided a more expansive list of twelve vineyards from which I could choose. Following are the names of the nominated vineyards.

Nominations for Famous and Historic Vineyards

Napa	Sonoma
• Beaulieu	• Bacigalupi
• Bosche	• Durell
• Diamond Creek	• Hanzell
• Eisele	• Hirsch
• FAY	• Maple
• Martha's Vineyard	• Monte Rosso
• Screaming Eagle	• Pagani
• Stag's Leap	• Rhinefarm
• Stagecoach	• Robert Young
• Three Palms	• Rochioli
• To Kalon	• Sangiacomo
• Winery Lake	• Seghesio Home Ranch

With this list as a guideline, I tried several methods to narrow it down to the ten most famous vineyards of Napa and Sonoma. The first was to pull all California wines scoring 98 points or above in *Wine Spectator, Wine Enthusiast*, and *Wine Advocate* magazines from 2000 to 2012. However, this method didn't work very well, because it became obvious that these three publications have a tendency to review different wine brands. Furthermore, the number of wines differed widely. For example, *Wine Spectator* gave 25 California wines a rating of 98 or above, and *Wine Enthusiast* awarded 42 wines, while *Wine Advocate* identified 371 California wines at this high level. Therefore, I abandoned this method.

The second method I tried seemed to work better. With this approach, I narrowed down the list by identifying vineyards that were located in a different AVA (American Viticulture Area) within the county and/or specialized in a separate type of grape varietal. In this way, a more diverse group of vineyards could be explored. Therefore, the final list of ten famous Napa and Sonoma vineyards for this book are listed in the table below.

Final Selection of Ten Famous Vineyards in Napa and Sonoma

Napa Vineyards & AVA	Sonoma Vineyards & AVA
• Beaulieu (Rutherford)	• Bacigalupi (Russian River)
• Diamond Creek (Diamond Mtn)	• Hanzell (Sonoma Valley)
• Stag's Leap (Stags Leap District)	• Hirsch (Ft Ross/Seaview)
• Stagecoach (Atlas Peak)	• Monte Rosso (Moon Mountain)
• To Kalon (Oakville)	• Seghesio (Alexander Valley)

AVAs in Napa and Sonoma Counties

Since this book focuses on ten famous vineyards in Napa and Sonoma counties, it is useful to provide a definition and list of the major AVAs in both counties. An AVA, or American Viticulture Area, can be defined as "a designated US grape-growing area, which is distinguished by unique geographical features." These usually include distinctive climate and geographical features and often a history of winemaking in the region.

AVAs must be approved by the US Alcohol and Tobacco Tax and Trade Bureau (TTB), and the process can take up to two years for approval. AVAs do not guarantee quality of wine but instead communicate an authentic and distinctive winegrowing region. Furthermore, there is no standard size for an AVA in terms of acreage. They can be quite small or span multiple states. They can also overlap one another.

In recent years the number of AVAs within Napa and Sonoma has been expanding. Following is a list of the majority of AVAs during the period in which this book was written. Napa refers to its AVAs as sub or nested AVAs within the valley.

AVAs of Napa and Sonoma

Napa AVA's	Sonoma AVA's
1. Atlas Peak	1. Alexander Valley
2. Calistoga	2. Bennett Valley
3. Chiles Valley District	3. Carneros
4. Coombsville	4. Chalk Hill
5. Diamond Mountain District	5. Dry Creek Valley
6. Howell Mountain	6. Ft Ross/Seaview
7. Los Carneros	7. Green Valley
8. Mount Veeder	8. Knights Valley
9. Oak Knoll District	9. Los Carneros
10. Oakville	10. Moon Mountain
11. Rutherford	11. Pine Mtn/Cloverdale Peak
12. Spring Mountain District	12. Rockpile
13. St. Helena	13. Russian River Valley
14. Stags Leap District	14. Sonoma Coast
15. Wild Horse Valley	15. Sonoma Mountain
16. Yountville	16. Sonoma Valley

AVA Map for Napa and Sonoma

It is also helpful to examine the AVA Map for Napa and Sonoma because it highlights how bodies of water, fog, temperature, and elevation impact the types of grapes grown in the various AVAs. For example, western Sonoma County is bordered by the Pacific Ocean, which causes this region to be much cooler and have more fog, a situation that is conducive to growing cooler-climate grapes such as Pinot Noir and Chardonnay. Likewise, the southern part of Napa County is bordered by San Pablo Bay, which creates a cooling effect on the Los Carneros AVA, where Napa grows most of its Pinot Noir and Chardonnay grapes. This AVA is also shared with Sonoma County and is home to many of the sparkling houses for both counties.

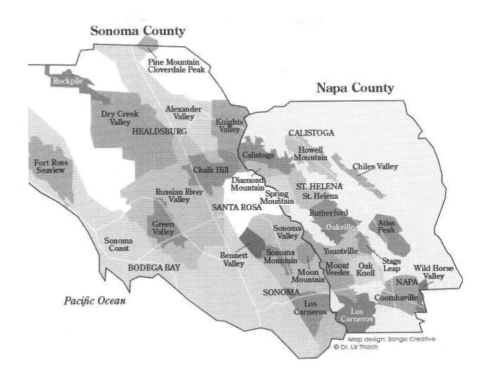

AVA Map for Sonoma and Napa

Driving north into the counties, the temperatures become warmer, and the mountain terrain becomes steeper. Therefore, in these areas grape varietals that prefer a hotter climate, such as Cabernet Sauvignon, Zinfandel, and Sauvignon Blanc, can be found. Often those grapes grown on the steeper hillsides tend to exhibit more intense flavors and structure than those grown in the more fertile valley floors. The moderating effects of the Russian and Napa rivers also impact the mesoclimate of vines grown near the water, providing more moisture in the soil and air.

Locations of the Ten Famous Vineyards

It is interesting to see how the locations of the ten famous vineyards fall within the Napa and Sonoma AVA map. Based on the set criteria, all of them fall into separate AVAs, with the exception of Monte Rosso and Hanzell, which are both within Moon Mountain AVA, but at this time Hanzell has elected to continue with Sonoma Valley on their labels. In addition, both vineyards have distinctively different soil types, microclimates, and signature grape varieties.

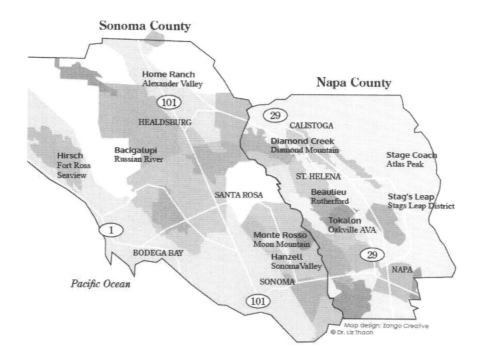

Location of Ten Famous Vineyards

Order of the Vineyards

Instead of dividing the vineyards into the separate counties of Napa and Sonoma, they are organized by season of the year. Not only does this highlight the beauty of the vines during their major cycles, but it signifies some of the unifying characteristics of the great vineyards that grace both valleys. The old feud of Napa verses Sonoma is passé, especially since so many wine companies are located in both valleys. In addition, wine tourism experts in both counties readily admit, "We are good for one another. We need and support each other."

Therefore the chapters of the ten famous vineyards are organized as follows:

- Winter - When the Vines are Sleeping (Monte Rosso, To Kalon)
- Early Spring – Bud break (Seghesio, Stag's Leap, Hirsch)
- Late Spring and Summer– (Stagecoach, Bacigalupi, Hanzell)
- Fall – Harvest and Autumn Leaves (Diamond Creek,h BV)

Chapter Two

Viticulture 101

As I was writing this book, I had several friends review the manuscript and then accuse me of throwing around strange viticulture terms they didn't understand. Therefore, I thought it would be useful to provide a quick overview of some of the terms and concepts that are so second nature to vineyard managers, they don't realize they appear to be speaking a foreign language to outsiders. For readers who already have a good understanding of viticulture, this chapter can be bypassed.

Vineyard Farming Methods

Vineyard owners use different farming methods around the world, and the choice of method is usually based on their personal philosophy of the best way to farm as well as cost issues. Following are the four most common methods:

1) **Conventional Viticulture**: Vineyard management techniques that are organized around a farming schedule with set timelines for irrigation, fertilizers, pesticides, and other conventional agrochemicals. In the past, most large commercial vineyards used this method, though many in California are now adopting sustainable viticulture.

2) **Sustainable Viticulture**: Vineyard management techniques that attempt to preserve the environment and are socially responsible, while

at the same time providing economic viability in the vineyard. In most cases, sustainable vineyards adopt Integrated Pest Management (IPM) techniques that encourage monitoring the vineyard to determine if water, fertilizer, pesticides, or other additions are needed, rather than administering them on a calendar basis. In many cases, sustainable vineyards use organic products but if necessary, resort to agrochemicals to protect the crop. There is also an emphasis on treating workers well and collaborating with the local community. Some vineyard regions have adopted codes of sustainable winegrowing to evaluate themselves and/or to become certified. The California Sustainable Winegrowing Program (SWP), Napa Green, and the Lodi Rules for Sustainable Winegrowing are good examples of such programs.

3) **Organic Viticulture**: Vineyard management techniques that do not include the use of any unnatural substance within the vineyard. Therefore, all conventional agrochemicals are banned from the vineyard, but natural substances such as sulfur can still be sprayed on the vines. According to USDA guidelines, a vineyard that produces "certified organically grown grapes" must be inspected by a government-approved certifier who inspects the vineyard to verify they have produced the grapes without using most conventional pesticides, sewage sludge, petroleum-based fertilizers, ionizing radiation, or bioengineering. Some growers choose to follow organic viticulture practices but don't get certified because of documentation and cost of doing so.

4) **Biodynamic Viticulture**: Based on the work of Rudolph Steiner, biodynamics views each vineyard as a living organism that can maintain itself if the soil and environment are nursed back to their natural condition before man intervened with chemicals and other unnatural systems. Viewed by some as an extreme form of viticulture, it involves farming according to the rhythms of the earth, such as planting new vines or pruning when the moon is descending. Biodynamics also includes preparing and administering field sprays made with cow manure, ground quartz, and herbs to bring the soil back into balance. Beneficial insects are encouraged to naturally assist in pollination and feed on predatory insects, as well as animals such as birds of prey to combat gofers. Additionally, chickens, cows, and sheep are often introduced to provide natural fertilization. Demeter Association, an independent international certifier of biodynamic farming practices, is usually called in to validate biodynamic practices.

Cycles of the Vine or a Year in a Vineyard

The cycles of a vineyard are a wondrous site to behold and can be quite motivational. Indeed, some of the great religions of the world use the analogy of a bare grapevine coming back to life after a winter of dormancy as a symbol of hope and renewal. I have a similar feeling about my small vineyard each spring when budbreak occurs. It is the signal of a new vintage and a new beginning. For many people in the wine industry, it is a time of celebration, just like the harvest season is another reason for festivity.

The cycles of the vine, which continue year after year, can be easily divided into the four seasons. Each season brings specific growth patterns in the vineyard, different types of work, and unique challenges.

Winter—Dormancy: The winter season, which generally runs from mid-November to mid-March in Napa and Sonoma, is a time when the vines are "sleeping." They are in a dormant state with no leaves or grape clusters. The work during this period is pruning of the vines and fertilization of the soil.

Spring—Budbreak/Bloom: Usually in mid-March, small, green leaves burst from the dormant black vines (budbreak), and for a brief week, the vineyard is a pale-green blur of new growth. Then the leaves and shoots grow rapidly. The work in the vineyard can be quite intense during the spring, because vines often need to be trained and tucked into trellis wires, extra shoots and leaves are removed (called *suckering*), and vineyards are usually sprayed with sulfur to prevent powdery mildew, which can attack during the damp spring weather. Other spring work may include mowing and discing cover crop between the rows.

In late April to mid-May, *bloom* occurs. This is where the tiny grape clusters develop small, white flowers, and for a brief few days, the vineyard is filled with a soft floral fragrance until the berries *throw the flower* and begin to grow into healthy clusters. The main threat in spring is frost that can damage the tender shoots, leaves, and clusters. Therefore, vineyard owners set frost alarms and often have to jump out of bed in the middle of the night to turn on frost turbines or sprinkler systems to protect the crop.

Summer—Verasion: Summer is a time of continued development in the vineyard, as the shoots grow even longer, and the grapes increase in size. From late June to mid-July, the grape clusters go through *verasion*, which is when they change color from green to red or purple. White grapes usually change from green to a paler green or yellow shade. Work in the vineyard includes continual pulling of leaves if they are shading the grape cluster too much and thinning of extra clusters, or *shoulders*, on clusters. This dropping of clusters is sometimes referred to as *green harvest*. The purpose of this process is to enable the vine to focus on ripening the primary clusters and not get distracted with too many extra clusters.

Summer dangers can include too much heat, which may sunburn or shrivel the grapes. Extra irrigation and proper canopy management to ensure there are enough leaves to shade the clusters are important preventives for this. In addition, as the grapes continue to ripen and become sweeter, they attract many animals desiring to eat them. In Napa and Sonoma, this may include birds, yellow jackets, raccoons, and deer. Bird netting, alarms, traps, and deer fencing are some of the methods that may be used to deter these hungry predators. Also, some vineyard owners put owl houses and bird of prey perches in the vineyard to encourage owls and hawks to manage the situation naturally, and they may introduce vineyard cats to patrol the vines.

Autumn—Harvest: In late August through early November, harvest commences in Napa and Sonoma counties. The timing is dependent upon the weather conditions for that vintage as well as the grape varietal. Early ripening varietals, such as Pinot Noir and Chardonnay, are often harvested in late August or early September, whereas late-ripening varietals, such as Zinfandel and Syrah, may not be picked until the end of October or even early November in some years. Cabernet Sauvignon and other Bordeaux varietals are often harvested in September and October. Grapes designated for sparkling wine, however, are collected first (usually in July), because winemakers prefer these grapes to have low sugar content, usually around 19° Brix.

The work during harvest is intense. Vineyard workers and winemakers often work around the clock to pick and deliver the grapes to the winery at the optimal time. Harvest generally takes place at night or in the early morning in order to keep sugar and acid levels at their prime. Workers may grab a nap in the late afternoon and then continue to work through the night. In some vineyards the work is done with mechanical harvesters if the terrain is not too steep, but operating the equipment and processing deliveries still requires workers.

The main danger at harvest time in California is a heavy rain. If this occurs, it can swell the grapes and dilute sugars, acids, tannins, and flavors. Generally this is not a problem in Napa and Sonoma, because the Mediterranean climate provides a warm, dry season from mid-May to mid-November and a cool, wet season the remainder of the year. This supplies the sunlight and warmth the grapes need to ripen in the summer, plus the cooling fog and lower temperatures at night to maintain acids and flavors, as well as plenty of rain in the winter and spring, so that in some years, irrigation is not needed.

Winter—Dormant Vines   *Spring—Budbreak and Bloom*

Summer—Growth and Verasion   *Autumn—Harvest*

Some Basic Vineyard Vocabulary

Following is a list of some of the common terms used in viticulture:

Rootstock: *Rootstock* refers to the root structure of a grapevine. Once upon a time, all grapevines grew on their own roots, but through the centuries they were attacked by root disease and pests that killed the vines. Indeed, *Phylloxera*—a small mite that feasts on the roots of vines—wiped out most of the vineyards of France in the eighteen hundreds and many in California in the 1990s. Therefore, the majority of grapevines around the world are now grafted to *rootstocks* that are more disease resistant. Some examples include St. George, 03916, 101-14, and 110R. In addition to protecting from some diseases, different types of *rootstock* can be more or less drought tolerant, and can also impact the vigor and quantity of fruit a grapevine produces. Each vineyard owner needs to analyze his or her soil to determine the best type of *rootstock* to use and in many cases will use a variety of *rootstocks* within the vineyard.

Clone: A *clone* is a population of grapevines derived from a single mother vine. Over time, grapevines *mutate* slightly to adapt to their specific environments. This results in the development of *clones*, which may impart a slightly different taste to the grape varietal, as compared to a different *clone*. *Varietal* refers to the name of the grape, such as Chardonnay or Cabernet Sauvignon.

Pinot Noir is well known for having many different *clones*, and some produce strawberry flavors, others produce mushroom notes, and some even produce smaller berries with more color. *Clones* are produced all over the world, with many famous ones coming from the Dijon and Pommard regions of France, but others have been developed in California, such as the Hanzell, Martini, Wente, and Swan *clones*. Vineyard managers usually plant a variety of *clones* of the same grape varietal because it adds more complexity to the final blend.

Vineyard Exposure and Spacing: When a vineyard is laid out, the viticulturist needs to determine which way the rows should run in order to get the best exposure to the sun. Too much sun can cause the grapes to be sunburned, whereas too little slows ripening. Therefore, exposure to sunlight is a very important consideration. For example, many of the most famous vineyards in Burgundy—*grands crus* vineyards—have an

Eastern exposure with the vines planted in the middle portions of the hillsides to achieve optimal sunlight and drainage.

Next, the spacing between rows and between vines needs to be determined. In Europe, they often prefer tight spacing of one meter across the row and one meter between the vines, or approximately three feet by three feet. In California, however, many vineyards are set up on wider spacing, such as twelve feet across and eight feet between vines, in order that large tractors can easily traverse the rows. Newer vineyards in California are now being planted with six by four feet and eight by six feet spacing, depending on the needs of the vineyard.

Vine Spacing in a 3 x 3 Foot Vineyard and a 6 x 4 Foot Vineyard

There is some debate over which type of spacing is best in producing the largest quantity of fruit, but in most cases it equals out. This is because if the vines are tightly spaced, they will compete with one another for water and nutrients, which results in a lower crop size; whereas if the vines are spaced widely apart, they don't compete as much and therefore will produce more fruit, but as there are fewer vines per acre, it usually ends up being similar in quantity. In terms of quality of fruit, this can often be determined through canopy management, where workers may pull leaves and drop excess fruit clusters in order that the vine can focus on ripening primary clusters to the highest level of quality.

Trellis Systems: In addition to spacing, the type of trellis system used in a vineyard can be quite important, as it determines how the vine shoots will grow, how much sunlight the clusters receive, and how much air penetrates through the canopy. All of these are important considerations depending on the climate. For example, in vineyards that receive a lot of sun and the grape clusters need to be protected from sunburn, trellis systems such as the California Sprawl, Lyre, Y-Shape, and Elkhorn may be used because they train the shoots and leaves down over the clusters. An opposite system is the Geneva Double Curtain that places the clusters toward the top of the trellis so they can receive more sunlight in the cooler climate of Geneva, New York, where it was developed.

Trellis systems also determine where the fruiting clusters will be located, which can impact how easily workers can pick the grapes. If trellis systems are low to the ground, workers may have to bend over a lot during harvest, which can create back safety issues.

Originally, grapevines used to grow along the ground and climb into trees and bushes, allowing their fruit clusters to ripen in a variety of locations. Over the centuries, however, vineyard workers trained them to grow upward around a stake in the ground into small, bush-like structures. Today, these are called bush vines or head-pruned vines in California. They can be pruned close to the ground to around three feet tall, or allowed to grow up to six feet tall like some of the historic Zinfandel vineyards of California. In France, a common derivation of this is the *goblet,* a short vine that grows up a stake to about two or three feet tall.

One of the most common trellis systems in the world is the VSP, or Vertical Shoot Position trellis. This simply trains the vine to grow straight (vertically) up a stake and then positions the shoots along a series of horizontal wires. It is a simple system to install and also allows for the grapes to be mechanically harvested. In France, a shorter version of this is the guyot where vines are trained up a shorter stake, and the shoots grow sideways along horizontal wires.

California Sprawl　　　　　　　　*Bush Vine*

VSP　　　　　　　　*Lyre*

Common Trellis Systems in California

Pruning: Regardless of the trellis system selected, the vines will also need to be pruned. There are two major types of pruning methods: *spur-pruned* and *cane-pruned*.

Spur pruned means the vine is pruned back to small *spurs*, each with two or three buds. Each bud will produce a shoot or cane that generally develops two clusters of grapes. *Spur pruning* is performed on *bush vines* or vines that have *cordons*—arms that extend horizontally to one or two sides. The number of spurs that are left each year depends on the size of the vine. For example, in California in vineyards on eight by six spacing with the VSP trellis system, vines with two *cordons* (arms) may be pruned to seven spurs per *cordon*, whereas in Bordeaux, where VSP *cordon*-vines are more tightly spaced, there may only be four to five spurs per *cordon*.

Cane pruned means that the vine is pruned all the way back each year, with the exception of several long canes, which are left to be tied to the wires. Each cane may have six to twelve buds, and each bud will produce a shoot with fruit clusters. *Cane pruning* usually produces a higher quantity of fruit, and is often used with cooler-climate grapes such as Pinot Noir and Chardonnay. However, it is also being adopted for other types of grapes, such as Cabernet Sauvignon in Napa, because cane pruning reduces the risk of *Eutypa* disease.

Soil Type and Irrigation Practices: Grapevines can be grown on many different types of soil, but in general, vines that produce the highest quality are produced from soil types that are well drained and are not overly fertile. In Europe, old-timers often say that vines should be planted on a slope in soil that will not support other crops. Indeed, this wisdom seems to play out in most of the high-quality vineyards of the world, including Napa and Sonoma, where some of the best fruit comes from hillside, or mountain vineyards on soil that can't grow much else. Soil composition, such as clay, limestone, and volcanic amalgamations, is also important, but a site where the vine struggles a bit and is forced to send its roots far below the surface to find water and nutrients seems to be the most beneficial to producing intensely flavored fruit.

Irrigation practices differ around the world, and in some parts of Europe, it is not legal to irrigate grapevines. However, in warmer climates, such as California, it is usually necessary to water vines in the hot summer months (though there are some people who disagree with this). A vine usually needs more water when it is first planted, and in the eighteen hundreds, this was done by hand with a bucket. Today most modern vineyards use drip irrigation systems that can be programmed to water vineyards when needed.

Technology in the Vineyard: Like many other industries, the wine industry has adopted technology to aid in efficiency, quality, and cost control. This also takes place in the vineyard where tools such as weather stations, soil moisture probes and satellite imaging are being used to determine irrigation, nutrient needs, and/or frost protection. The ten vineyards described in this book use a variety of technology that is explained by each vineyard manager.

Canopy Management: refers to the methods used to manage the number of leaves and shoots (the canopy) on a vine. Left to itself, a grapevine will grow many more leaves and shoots than are needed to ripen its fruit clusters and in many cases will not ripen the fruit to the level needed for quality winemaking. Therefore, vineyard managers will monitor the canopy and remove any leaves, shoots, or extra clusters that are taking away vital carbohydrates from the primary clusters. Canopy management also refers to the majority of the items described previously, including spacing, trellis system, pruning, irrigation, and technology in the vineyard.

Pest & Disease Control – Two other viticulture terms that are often mentioned are *pest control* and *disease control*. Vineyard managers spend a lot of time monitoring the vineyard to protect it from these issues, but they do vary by country and region. For example common *pests* in the vineyards of California may include gophers, mites, and yellow jackets. Occasionally, if deer-fencing is not used, deer may enter the vineyard and eat the grapes. Interestingly in Europe wild boar are often a pest control issue, whereas Australia may have kangaroos in the vineyard and South Africa has reported issues with baboons eating their grapes.

Disease control is usually more problematic than pest control. The most common disease in vineyards around the world, including California, is *powdery mildew*. It is a fungal disease that can result in loss of yield and reduce wine quality. Other diseases include *grapevine leafroll, redleaf, Pierce's disease, downy mildew, phylloxera, Eutypa*, and many others. The ten vineyard managers in this book describe pests and disease issues they encounter and how they are controlling them.

Important Measurements of Grape Ripeness

Harvesting wine grapes at the correct time is one of the most important precursors of making quality wine. Without natural flavors, sugars, acids, and tannins at the ideal levels, the wine will not taste as flavorful, be well balanced, or exhibit good intensity and length. Of course, a winemaker can add or adjust some of these components at the winery if the measurements are not correct by adding acid, for example, but most winemakers prefer to obtain these elements naturally through quality wine-grape growing.

There are five key elements that a winemaker considers when making a decision when to harvest grapes. These are: *Brix, acid, pH, tannin structure*, and *flavor profile*. Though most top winemakers can walk through a vineyard and determine by taste alone whether the grapes are ready or not, the majority still also use laboratory measurements as verification.

1) **Brix** identifies the level of sugar within the grapes and determines the alcohol potential. The higher the Brix, the more sugar the grape has and the higher the alcohol may be in the wine. For example, a Brix of 23° generally indicates the wine will have a 12–12.5 per cent alcohol, whereas a Brix of 26° suggests it can climb as high as 14 percent or more. In general, winemakers are looking for a Brix level between 22° and 26° for a still wine. In other parts of the world, different measurements for grape sugar are used, such as degrees Baume and Oechsle.

2) **Acid** determines how acidic the wine may be. If the grapes have too much acid, the wine may taste sour on the palate. If the acid is too low, the wine will taste flabby. When measuring acid in wine, labs generally analyze titratable acidity, with the acid level for white wines ranging from 7.0 to 9.0 and for red wines, 6.0 to 8.0.

3) **pH** indicates the potential of hydrogen ions in the wine and is a measure of the relative acidity in a wine. A lower pH reading indicates higher acid strength in the wine, whereas a higher pH shows lower relative acidity. In general, winemakers try to achieve a pH range of 3.0–3.4 for white wines and 3.3–3.6 for red wines.

4) **Tannin** indicates the level and structure of the tannins in a wine. Tannins are derived from the grape skins, grape seeds, and also the oak barrels used for aging. Generally in the vineyard, winemakers just focus on tannin structure for red wines. Though this can be measured, they often determine tannin structure by tasting the grape and looking at the color of the seeds. If the grapes still seem bitter and rough in texture and the seeds are green, the tannins are most likely not ready.

5) **Flavor Profile** indicates the taste of the fruit flavors in the grape. Winemakers taste the grapes to see how well the flavors have developed and if they need more hang time to develop deeper and richer fruit

flavors. Also, if the grapes still taste *green* or with vegetative notes, they are generally not ready to be harvested.

Grape Tonnage and Case Production

A common question asked of California vineyard managers is, "How many tons per acre do you get?" That is because *tons per acre* is the method to determine production yields within a US vineyard. A ton is two thousand pounds of grapes. Therefore, if a vineyard produces an average of two tons per acre, that means they produce four thousand pounds of grapes per acre.

Another guideline is that, in general, you can produce around *800 bottles of wine or 160 gallons from 1 ton of grapes*. This assumes an average amount of juice pressed from the grapes. Some winemakers can press more juice from a ton of grapes to make a mass-market wine, whereas high-end wineries may only choose to use *free-run juice*, which results in much less wine. *Free-run juice* is the wine that flows freely after fermentation and is not pressed. In general, it is considered to be higher quality than heavily pressed wine.

Tonnage per acre ranges widely in California vineyards, with some vineyards in the Central Valley easily achieving sixteen to twenty tons per acre. However, most high-quality vineyards in Napa and Sonoma range from one to four tons per acre. This is because these vineyards are carefully tended, and clusters are dropped so the vine can ripen the remaining clusters to the perfect level of flavor, Brix, acid, pH, and tannin. Therefore, tonnage per acre on high-quality vineyards is usually lower.

In Europe and South America, production is measured in *hectoliters per hectare*. In general, the conversion for *1 ton per acre in the United States is equivalent to 13.5 hectoliters per hectare* in Europe. For example, in the Medoc AOC of Bordeaux, regulations require that the maximum yield a vineyard can produce is 50 hectoliters per hectare. However, many of the grands-crus wineries harvest much less, with yields ranging from 20 to 27 hectoliters per hectare, or 1.5 to 2 tons per acre.

Finally, in the United States, wine production is generally measured in *number of cases produced*, whereas in Europe and other countries, wine production is often quantified as *total number of bottles produced*. Since there are twelve bottles in a case, when converting European wine production, simply divide by twelve. Therefore, a Burgundy winery that produces sixty thousand bottles per year would be equivalent to a five-thousand-case winery in the United States.

A Few Words on the Concept of *Terroir*

Whenever a group of wine enthusiasts come together, the term *terroir* is often bandied around. It can be a controversial term, and some experts deem the concept as hogwash, whereas others profess it to be the soul of a great wine. I happen to fall in the second camp, especially after planting and nurturing my own hobby vineyard for the past ten years.

Terroir can actually be defined at three levels. The first is soil or the taste of the dirt, which is part of the literal meaning of the term *terroir*. The second and most commonly used definition is that *terroir* is the combination of soil, climate, water, sun, and topography (altitude and slope). However, I prefer the third and more holistic definition that encompasses all of the above and adds farming and winemaking practices to it. This is because decisions made on how to farm the vineyard, as well as regional regulations on what is acceptable, such as amount of irrigation, can impact the taste of the grapes upon harvest.

Furthermore, most winemakers are intimately involved with vineyard farming practices and have opinions on how many clusters should be left on the vine to ripen each year and when to harvest. Additionally, when the grapes arrive at the winery, the winemaker determines the degree of sorting that is necessary, as well as the type of yeast to use, and any other needed additions that the vineyard may not have provided that year. Therefore, the holistic definition of *terroir* is more valid in determining the final taste of the wine.

The advantage of this definition is that it recognizes all of the variables and resulting complexities of crafting a fine wine. The hands of mother nature are joined by the hands of the viticulturist and winemaker to create a unique product that is a holistic expression of place.

Searching for the Soul of the Vineyard—Different Perspectives

There are different perspectives on the best way to grow wine grapes. Some of my friends and acquaintances recommend treating a vineyard like a big garden, whereas others are more philosophical and find deeper meaning in the coexistence of vineyard, animals, and humans.

According to one pragmatic friend, "It's no big deal. Tending a hobby vineyard is just like taking care of a garden. If it needs water, then water it. If you see it's getting powdery mildew, spray with sulfur or something else. If the birds are eating your grapes, then put up nets. No need to stress over it."

This was comforting advice when I first planted my hobby vineyard. I have to admit that I was rather worried, because I'm not very good at even keeping houseplants alive. My schedule is so busy that I forget to water and fertilize them. Instead, I have many silk plants in my house.

"A grapevine is very forgiving," he continued. "If you happen to prune it incorrectly this year, there's always next year. They are very hardy plants."

I have to agree that they are hardy in that the foliage comes back every year in radiant green. However, the fruit may not be as plentiful or of the quality you desire if you don't take the time to nurture the vines correctly.

And then there is the more expansive viewpoint of experts like Paul Dolan, who has rightly been called "The Godfather of Sustainable Viticulture" in California and likens a vineyard to a living organism.

"It has a circulation system, which is the water flowing from the mountains into the soil and the vines. It has a respiration system in that the wind and air circulate around the leaves and clusters. It has a digestive system because animals eat the grass in the vineyard, and we use their manure to create fertilizer that is put back in vineyards to nurture the vines."

Paul farms his Dark Horse Vineyard in Mendocino County using biodynamic principles and practices. He continues by describing why he adheres to this philosophy. "Most farming is exploitive in the sense that we remove nutrients each time we harvest, so we like to regenerate the vineyard using only organic materials. Additionally we recognize that there is also a loss of energy at harvest, which we replenish using biodynamic practices. All of this is in service of contributing to the overall health of the space in which the vines grow such that the vines have the ability to more fully express themselves both in the fruit and the wines".

As I consider my small hobby vineyard, Paul's biodynamic vineyard, and the hundreds of other vineyards I have visited around the world, I realize they all have one goal in common, and that is to produce wine grapes. Whether these grapes are destined for jug wine, mass market, luxury wine, or somewhere in between, the vineyards are all farmed to produce. But it is the intention behind the farming methods, as well as the location, history, and regulations of the region that determine the soul of the vineyard. Therefore, in the rest of this book, we will examine ten famous vineyards of Napa and Sonoma and try to determine what makes them so special and what they can teach us.

Winter

"When the Vines are Sleeping"

Monte Rosso Vineyard

To Kalon Vineyard

Chapter Three

Monte Rosso Vineyard
Moon Mountain AVA, Sonoma County

It was a perfect blue-sky day in December as I drove the back roads of Sonoma County to Monte Rosso Vineyard. The air temperature gauge on my car read 42° F, but the sun shone brightly on the green hills and sculptured vineyards I passed along Bennett Valley Road through Glen Ellen and then onto Highway 12, the main artery connecting Santa Rosa to Sonoma. December is a month when there are fewer tourists in wine country, but the beauty of the vineyards continues to be splendid, as the vines show off their dark limbs clearly in the sunlight, and a few leaves of yellow, gold, and orange still cling to shoots.

The driving directions to Monte Rosso were fairly simple, and I was feeling confident about arriving on time for my ten o'clock appointment until my GPS said, "Make a U-turn and go back." Apparently I had missed the tiny sign signaling the approach of Moon Mountain Road. Turning my SUV around, I headed up the twisting road to the famous vineyard, passing oak trees shielding charming neighborhood houses, which became swankier the higher up the hill I climbed.

Eventually, an impressive sign stating Monte Rosso Vineyard appeared on the right, and I followed a small paved road into a broad valley of undulating hills covered with expansive vineyard blocks. Some were gnarled, old bush vines that rose tall and proud with their curving black limbs outstretched. Others were newer vines on orderly trellis systems, marching like soldiers into the distance. I stopped the vehicle several times to snap photos because the sunlight was brilliant on the bright-green grass and the remnants of orange leaves and small purple clusters of a second crop on the black vines.

Winter Vines in Monte Rosso Vineyard

After driving about five minutes through multiple vineyard blocks and wondering how someone had the foresight in the 1880s to clear this mountaintop of rocks and scrub oak to plant this expansive vineyard, I came to a group of buildings, including a barn, house, several outbuildings, and an old 1940s vintage truck. Standing in the dirt driveway was Jake Terrell, vineyard manager for Monte Rosso but using the official title of Ranch Leader. He was accompanied by Kelly and Colleen, two public relations representatives from Gallo, the current owner of Monte Rosso.

Jake was wearing jeans and a black down jacket as he approached to shake hands, and introductions were made all around.

"Welcome to Monte Rosso," he said with a quick smile. His short, blond beard, mustache, and hair glinted red in the sunlight, and I learned shortly that he also had a wry sense of humor. Jake graduated from Cal Poly's Agribusiness program, and after working at Justin Vineyards and Winery in Paso Robles for a while, he took a position with E&J Gallo in Paso Robles for four years before being promoted to vineyard manager for Monte Rosso. When I asked what compelled him to study viticulture, he replied, "I've always loved science, and to me, farming is applied science."

The four of us climbed into Jake's large 4-door truck, while his hound dog, Willie, jumped in the back, and we headed into the vineyards. Within a few minutes, a coyote appeared, running through a block of vines in front of us.

"I hope Willie doesn't see the coyote," said Jake, "otherwise he will chase it." But Willie was looking the other way in the back of the pickup, and when we stopped, he jumped out to chase a rabbit in the opposite direction.

Jake smiled wryly. "Well, I guess I don't have to call the boys to cue the animals to appear for you," he said jokingly. "I was going to ask them to bring in a giraffe, but as you can see, the vineyard has plenty of native wildlife of its own."

History of Monte Rosso Vineyard

Monte Rosso is one of the oldest vineyards in California. The term *Monte Rosso* means red mountain because of its location on the top of the Mayacamas Mountains with their distinctive, red soil. The vineyard was established by Emanuel Goldstein in 1880 when he bought the property with his partner, Ben Dreyfus. Their original name for the site was Mt. Pisgah, which is a Hebrew term for high place.

Emanuel and Ben hired a group of Chinese laborers who spent months clearing the vineyard site of brush and rocks, and then they planted the first vines in the early 1880s. Unfortunately in 1886, Ben passed away, but Emanuel kept the vineyard operating, and also built a stone gravity flow winery, which became known as the Goldstein property.

In the 1890's he replanted several of the old vine zinfandel blocks that are still in existence today. He expanded the winery to produce 20,000 cases, and operated it successfully until Prohibition when the government forced him to shut it down. Emanuel managed to stay in business by maintaining the vineyard and selling grapes to home winemakers.

After Prohibition, Emanuel continued selling grapes commercially but never restarted the winery. One of his clients turned out to be Louis M. Martini, who was very pleased with the high quality and taste of the grapes from the Goldstein property, especially the old vine Zinfandel. Therefore, when Emanuel died and Louis learned that Emanuel's heirs planned to sell the property, he contacted them in 1938 and was able to purchase the entire 550-acre property, including the 180 acres of vineyards.

One of the first actions Louis took was to rename the vineyard Monte Rosso, in honor of its beautiful red soil. Louis also planted several blocks of Cabernet Sauvignon that first year, and today these sections of the vineyard are considered to be some of the most historic old-vine Cabernet Sauvignon in the United States. By the 1940s Louis had increased the vineyard plantings to 250 acres and used surplus World War II army vehicles to help farm the steep hillsides.

Red Soil of Monte Rosso Vineyard

For many decades Louis and his family used Monte Rosso fruit to make wine in his famous Napa Valley winery, Louis M. Martini, and also sold excess fruit to other winemakers. Louis passed away in 1998, leaving his winery and the Monte Rosso vineyard to his heirs. His son, Mike Martini, became head winemaker. When I called him to arrange for an interview, Mike recalled many wonderful experiences working in the Monte Rosso vineyard.

"The vineyard means a lot to me," he reported. "I grew up playing among the vines and even proposed to my wife at sunset one evening in the Monte Rosso vineyard."

In 2002, the Louis M. Martini family sold the winery and Monte Rosso Vineyard to E&J Gallo.

"The Gallos and the Martinis have been friends for generations," explained Mike. "It has been a good partnership, because we are both old Italian families in the wine business."

Today, the largest percentage of Monte Rosso fruit is designated for Louis M. Martini wines, and the Martini family still maintains control of how the fruit is sold. The remainder goes to eighteen other lucky wineries that have been fortunate to secure grape contracts with Monte Rosso, including such esteemed names as Arrowood, Sbragia Family Vineyards, Rosenblum, Bedrock, and Biale wineries, among others.

Touring Monte Rosso Vineyard

"The total property is five hundred and seventy-five acres, of which two hundred and twenty-five are currently planted to vineyards," reported Jake as he continued to steer the truck up a winding dirt road toward the top of the mountain. We passed a hillside that was covered with green grass and terraces but had no vines.

"We are getting ready to replant that block," said Jake. "It is a lot of work to survey and plant on these steep curving rows, but the results should be awesome. We have left the land fallow for five years to allow the soil to rest and regenerate before replanting, as part of our environmental practices."

We reached the top of the mountain, and Jake parked the truck next to a spectacular head-pruned vineyard of gnarly old vines with huge trunks and thick curling arms.

"We have fifty acres of old vines Zinfandel, with several blocks dating back to 1890," said Jake. "This is the Rattlesnake Block. On hot days, we occasionally find a rattlesnake curled up the vines, but it is too cold today," he said reassuringly.

The view from the top of the mountain across the vineyard was stunning. I looked around and could see a vast panorama of undulating vineyards, hills, and broad sky in all directions. It felt as if I was standing on top of the world. In the distance I could see the glistening waters of San Pablo Bay.

Old Vine Zinfandel in Monte Rosso Vineyard

"If you look over there, you can just make out the top of the Golden Gate Bridge," said Jake.

I squinted my eyes, and sure enough, there was a hazy outline of the famous bridge.

We walked into the vineyard, and I reached out to touch one of the gnarled old vines. The bark felt smooth and slightly shaggy beneath my fingers. A few stray orange leaves clung to the limbs, and slightly shriveled clusters of second crop hung from some shoots.

"Taste them," said Jake as he popped a berry into his mouth.

I reached out and gingerly pulled off a soft, purple grape. It burst sweet and jammy on my tongue.

"You could make late-harvest Zinfandel with these," I said.

"Some years we invite neighbors in to harvest the second crop," responded Jake.

Back in the truck, the next stop was the Los Ninos Vineyard, which is one of the Cabernet Sauvignon blocks planted by Louis M. Martini in 1938. The vines are situated on a gradual slope on the side of the mountain, and as we walked through the rows, I was mesmerized by their ancient beauty. The spacing was wide, at eight by eight feet, and they were all *head pruned* with massive black trunks and curling arms. Most striking were the bright-red leaves with yellow veins that still clung to the vines, interspersed with small, dried, purple clusters of second-crop fruit.

"As you can see," said Jake, "these vines have leafroll virus, which causes the leaves to turn that bright-red color, but the virus doesn't hurt the quality of the fruit. However, it does decrease the production. We only get about two tons per acre from this block, but it makes one amazing wine!"

Jake reached out and touched one of the big red leaves. "Some growers would recommend pulling out these vines," he said, "but here at Monte Rosso, we rarely take out old vines. Our philosophy is to preserve them because of their history and our belief that they produce unique flavors even though they yield less tonnage than the new blocks."

We continued our drive across the side of the mountain, passing many other blocks of vines. Some were newer with slender trunks trained on

VSP (vertical shoot position) trellising with tighter spacing of around eight by five, where others were obviously older with thicker trunks and wider spacing.

We stopped at the Gnarly Vine Block, which was an impressive five acres of 115-year-old Zinfandel. I gazed out across that vast army of gnarly old soldiers with their twisting black arms and realized they had stood in this location since 1890—living through two world wars and into the Information Age. In addition, they had probably produced enough fruit to make thousands of bottles of wine.

Jake and Cabernet Sauvignon Vines in Monte Rosso

"Now we're going to see the Thieves," Jake interrupted my reverie.

"The Thieves?"

"Yes," he said as he started the truck. "The workers who prune this block named it the Thieves, because the canes are so long they get tangled up around your legs. They say the canes steal things out of their pockets when they walk by."

As soon as I saw the block, I could understand why. This was an ancient grouping of 3.8 acres of Semillon vines, also planted in 1890. They stood on very wide, fat trunks with long, bare canes protruding in all directions, like a group of giant octopuses.

"Semillon," I said, surprised. "They planted Semillon here in the 1890s? Were they trying to make Sauternes?" The traditional sweet French dessert wine is usually made of *botrytised* Semillon and Sauvignon Blanc grapes.

"Not sure," said Jake, "but one of the winemakers who buys this fruit has made a sweet dessert wine from it. However, most generally make a dry white Semillon. It is quite tasty."

I gazed at this amazing, tangled vineyard and thought about how complicated it would be to prune those long shoots every year. But the Thieves sat there unperturbed, dozing in the sun, and looking forward to the opportunity to steal more items from the pockets of unsuspecting pruners.

As we headed back down the mountain toward the main cluster of vineyard headquarters buildings, we came to a tree-shaded section where a small stream sparkled next to an old stone building. Jake idled the truck so we could look out the window.

"That is the original winery that Emanuel Goldstein built here in 1886," he announced. "Currently the building is empty, but it might be fun to retrofit it someday."

I looked at the trees shading the ancient stone building with the stream running close by and realized what an ideal location this was for a winery in the eighteen hundreds. The area was naturally cooler, which would help to protect the fermenting and aging wine during a time when there was no air conditioning as there is in today's modern cellars. The old winery, with its massive gray stone walls, stood two stories tall and seemed to whisper with secrets of bygone days when workers converted tons of grapes into dark red wine within its cool, dark interior.

Vineyard Specifics

Throughout our tour, Jake described some of the specific technical details of the vineyard. In terms of varietals, in addition to the Zinfandel, Cabernet Sauvignon, and Semillon, Monte Rosso has plantings of Merlot, Malbec, Petite Verdot, Cabernet Franc, Syrah, and Petite Syrah.

"We also have some Folle Blanche," announced Jake. "As you probably know, it is a French grape traditionally used to make Armagnac."

I asked if any of the old Zinfandel blocks were *field blends*, which is a term used to describe the traditional planting method used by many Italian immigrants. Instead of making separate batches of wine in the cellar and blending the different grapes to make a finished wine, the old Italians used to "blend in the field" by planting different varietals intermixed into the same vineyard.

"Yes, indeed," responded Jake with enthusiasm. "We've had some analysis of the old Zinfandel blocks and have discovered that they range from eighty to ninety-nine percent Zinfandel with the remainder being mixed blacks."

"Mixed blacks?" I asked, puzzled.

"Yes. It is a term for some of the old European grapes that actually produce dark juice, rather than white juice, like classic red grapes. Examples include Alicante Bouschet and Grand Noir."

"Fascinating!"

The distinctive, red soil of Monte Rosso is composed primarily of Red Hill loam and decomposed volcanic ash. Jake bent down to pick up a handful of the muted red-orange dirt.

Vineyard Specifics for Monte Rosso

Total Vineyard Acres	225 acres
Varietals	Zinfandel, Cabernet Sauvignon and Semillon, Merlot, Malbec, Petite Verdot, Cabernet Franc, Syrah, Petite Sirah, Folle Blanche, Alicante Boushet, Grand Noir
Soil	Red Hill loam, decomposed volcanic ash
Elevation	1000 to 1250 feet
Average Temperature	Summer 85F day with 55F at night
Rootstocks	St. George, Lenoir, 1103,110R
Clones	Zinfandel = Monte Rosso selection Cabernet Sauvignon = original from Flora Springs, 4, 8, 357
Sun Exposure	South
Spacing	Old Vines = 8 x 8; New Plantings = 10 x 4 or 8 x 5
Trellis Systems	Bush vine, 2 wire California sprawl, Lyre

"The topsoil is eighteen to twenty-four inches deep," he stated, "and is composed of clay and loam. Beneath the topsoil is fractured rock, which provides excellent drainage for the vineyard. Our pH is a little low, so we sometimes add some lime to our compost. Also, when we install a new vineyard and run soil samples, we usually find we need to add some lime to the soil and disk it in. However, this is actually a positive because Monte Rosso is known for producing naturally high-acid wines, and it is probable that the soil plays a part in that."

As it is situated in a valley high atop the Mayacamas Mountains, the elevation of Monte Rosso ranges from 1,000 to 1,250 feet at the top. The vineyard is classified as Region II, which is cooler than Napa, with summer temperatures averaging 85° F in the daytime and lows of 55° F at night. Average rainfall is around thirty inches per year but varies by vintage.

"Due to its high elevation," reported Jake, "a unique aspect of the vineyard is that it receives plenty of sunlight and never has frost."

Because the original 1880 plantings of Monte Rosso fell victim to *Phylloxera*, they were replanted in 1890 on St. George and Lenoir *rootstock.*

"The *clone* for the Zinfandel vine is what we call Monte Rosso Selection," explained Jake. "No one really knows where it came from originally, but it produces large clusters. Like most Zinfandel, it also ripens unevenly, but it is warm enough here that we usually have it all harvested by the end of September.

"The old block of Cabernet Sauvignon is originally from Flora Springs in Napa, but we're not sure which *clone* it is. The new cab blocks have been planted to a variety of *clones*, including 4, 8, and 357, and we use modern selections of *rootstock*, such as 1103 and 110R, depending on location and soil."

The main *sun exposure* on the vineyard is south, but there are a few blocks that run in different directions. In terms of *spacing and trellis systems*, the old vines are primarily on eight by eight spacing and are *head-pruned* with no trellising. For the new blocks, Monte Rosso mainly uses ten by four spacing, but density is based on the soil type. Where the soil is richer in nutrients, they increase spacing to encourage competition. A majority of the newer trellis systems are a two-wire California sprawl, but they are now installing Lyre trellising in some of the latest blocks.

"I like the Lyre," explained Jake, "because it provides good spacing for fruit and decent light intercept."

All vines in Monte Rosso are *hand pruned.*

"Most blocks are *spur pruned*," said Jake, "with the exception of the Semillon, where we do *cane pruning.*"

Farming Practices—Certified Sustainable and ISO 14001

When we reached the topic of farming practices and philosophy, Jake was obviously in his element. "I love to talk about the nerdy side of farming," he smiled.

Kelly, the PR Manager with E&J Gallo, also had much to offer. "Julio Gallo always had the philosophy of fifty percent/fifty percent when it came to vineyard operations," said Kelly. "Therefore, wherever Gallo operates a vineyard, they only plant on fifty percent of the land and leave the rest natural. It is part of our support and regard for the environment."

Another example of their regard for the environment is the fact that Monte Rosso is a *certified* sustainable vineyard with both the California Sustainable Winegrowing Alliance and ISO 14001. This includes certification of all vineyard operations for both environmental and management programs with employees and the community.

"In terms of *fertilization*," explained Jake, "as a sustainably certified vineyard, we always try to use organic products if possible."

Green cover crop is planted immediately after harvest. This includes a mix of grass, barley, bell beans, and clover. In May, compost is deposited over the cover crop, and then it is disked into the soil. The compost is usually manure and other natural ingredients. *Petiole samples* are taken every spring to analyze vine nutrition, and if necessary, natural supplements are added to the soil. *Weed control* is managed primarily through mowing.

The day-to-day work in the vineyard is busy, with canopy management, including suckering and thinning, a continuing task throughout the growing season.

"By the time we finish suckering," Jake reported, "it is time to begin leaf pulling. This is important in this vineyard in order to control powdery mildew and ensure the clusters have dappled light. Obviously, we consider the exposure and may take more leaves off one side of the row than the other. The Zinfandel is the most challenging because we have to pull more leaves on Zinfandel in order for it to get ripe."

At Monte Rosso they also utilize the practice of green harvesting, which means dropping or thinning extra grape clusters, so that the vine can focus on ripening its main clusters.

Farming Practices at Monte Rosso

Certifications	ISO140o1, Certified California Sustainable Winegrower
Fertilization	Green cover crop, natural compost, natural supplements
Weed Control	Mowing
Canopy Management	Suckering, thinning, green harvest of excess bunches and shoulders
Disease Control	Organic sulfur to prevent powdery mildew; helicopter occasionally
Pest Control	Mites main issue – water road to reduce dust that attracts mites
Irrigation	Drip, and dry farmed for old vines
Technology	Weather station, NDVI Maps
Harvest Measurements	Winemakers usually determine based on taste

"We drop fruit right after fruit set," said Jake, "usually taking off the wings and any extra clusters. On the old bush vines, ideally we want the clusters to look like well-balanced Christmas ornaments. We will go through the rows several times to drop fruit as needed for that year."

In terms of *disease control*, the main issue they have to watch for is powdery mildew, for which they apply organic sulfur every seven to fourteen days as a preventive measure.

"We look at the UC-Davis PM Index (a tool that helps determine powdery mildew growth based on canopy temperatures) and determine when we should do an application," said Jake. "Later in the season, when we can't get the tractors between the *head-pruned* vines because the canes are so long, we will do helicopter spraying with organic sulfur— usually early in the morning."

At Monte Rosso, they also scout the vineyard on a weekly basis to identify any other *pest control* or disease issues. Mites, which are tiny bugs that damage grape leaves, are occasionally a problem. Therefore, the vineyard makes sure to keep dust to a minimum on the roads, because dust can attract mites. Other potential vineyard pests such as deer, birds, gophers, yellow jackets, and raccoons, are not much of a problem, according to Jake.

"The property has deer fencing, so they cannot get in to eat the leaves. Gophers usually only attack young vines, so we carefully monitor new plantings," he said. "But in general, birds and animals are not a problem at Monte Rosso."

Drip irrigation is used on all new blocks and along some of the old bush vines if they have had to replant one.

"Lyre trellising needs more water than VSP," stated Jake, "but we only use irrigation if it is needed."

In terms of other *vineyard technology*, they have one weather station and also utilize NDVI Maps (Normalized Difference Vegetation Index).

"Each year we hire a pilot and plane to fly over the vineyard and take photos," explained Jake. "These are analyzed so we can see on a computer screen which sections of the vineyard may need more water or fertilizer. In my opinion, the key for quality fruit is uniformity. That's why we want to have a big-picture overview of the total health of the vineyard."

As we moved toward the conclusion of the tour, I asked Jake what types of harvest measurements are used to determine the correct time to pick the grapes.

"The winemakers who have contracts on the different blocks usually just come and taste the grapes to determine when they want to harvest," responded Jake. "Most of them don't use numbers to decide."

Economics of the Vineyard

Overall, Monte Rosso has an approximate *average yield* of two to three tons per acre.

"The old vines," stated Jake, "average between one and two tons per acre, depending on the vintage, whereas the newer blocks produce three to four tons per acre." Therefore, the production for all 225 acres would average around 788 tons per year.

In terms of *farming costs per acre*, I was told this information was not available; however, employee numbers could be shared. Gallo's Sonoma Ranches are all union properties with negotiated hourly pay and benefits for workers. At Monte Rosso, there are nine full-time workers who live on the property, but during harvest the work force can swell as large as fifty people. Safety is very important on the ranch, and they have adopted a behavior-based safety program.

"The workers take turns observing one another in a safety audit," reported Jake.

Regarding vineyard *revenues*, prices per ton are negotiated on an annual basis with all of the wineries that purchase fruit from Monte Rosso. Though they couldn't reveal an average price per ton, Jake and Kelly could verify that "All of Monte Rosso fruit sells for above the average price of Sonoma County fruit." The 2013 price per ton for Sonoma grapes averaged $2,235, according to the *2013 California Grape Crush Report*.

Economic Viability of Monte Rosso Vineyard

Average Yield	Approximately 2 to 3 tons per acre: old vines 1 to 2 tons per acre and new vines 3 to 4 tons per acre
Total Average Tons Per Year	788
Costs Per Acre	Not Available; union property
Revenues	Above average Sonoma County prices of $2235 per ton – negotiated by individual contract and block
Economic Health	Very good; waitlist to buy grapes

In the end, when I asked about the overall *economic health* of the vineyard, Jake responded, "I am responsible for the budget, and the vineyard has very good economic viability."

The Soul of the Vineyard—"Magical"

As Jake steered the big truck back into the dirt parking lot of Monte Rosso vineyard headquarters, I felt rather sad that the tour was ending. We climbed down from the truck, and I rubbed my hands together as I realized the day was still quite cold, hovering in the low fifties. Looking around the vast property, I asked my last few questions intended to discover the soul of the place.

"Jake, if you had to describe the essence of the vineyard in one word, what would it be?"

"If I had to use one word to describe Monte Rosso," mused Jake, glancing around at the nearby vines, "I would say magical. It's not just that it is a beautiful place to work, but I keep discovering unique things about this property. For example, one day I found some of the old bridles that were used on the horses that plowed the vineyards over a hundred years ago."

Jake turned and gestured to an old, rusted plow sitting on the dirt driveway a short distance from the barn. "And another day, I found this plow buried beneath an old blackberry bush. I also discovered an old distillery, a stone statue of Buddha, and the original bonded-winery sign buried in the dirt floor of the cellar building."

Kelly smiled and said, "It's almost like Monte Rosso has its own spirituality."

Jake nodded a silent yes. "And we all recognize that every vine is different. I have one employee who is like an artist when he prunes the vines. Every cut is perfect, and you can tell he loves what he does. All the guys who work here realize how special this vineyard is. They have a strong sense of ownership."

I asked Jake what he liked best about working at Monte Rosso and to describe some of the challenges of working such a vast vineyard.

"What I like best about working here," he responded, "is that I am one of the caretakers of Monte Rosso. This is not just something to put on my resumé but a great honor that feels cool and brings much credibility." He paused and looked down at the ground. "In fact, recently I was introduced to Jesus, the former vineyard manager who was here for twenty years and retired before I arrived. Even though we had just met, we talked for over an hour, and I asked him so many questions. Then I invited him to come back and visit anytime, but he shook his head and said, 'I can't. It will probably make me cry, because I miss it.'" Jake turned and stared directly at me. "So you see, there is a strong emotional connection here for many of the workers."

Old Monte Rosso Winery Building

In terms of challenges, he described how complicated it can be at times to keep track of sixty very different blocks of vines.

"Each is unique," he said, "and there is much complexity to deal with, as well as a lot of paper work. Other challenges can sometimes be the interactions with the twenty wineries that buy grapes from us. We negotiate contracts every year, and most of them are great to work with, but there have been a few that are more difficult."

The final question had to do with key learnings from working the vineyard. Jake responded readily.

"What I've learned from the vineyard is how little impact we have on quality. True grape quality really comes down to *terroir*. In my opinion, site and vintage triumph everything, and a number I would use for this is ninety percent. As vineyard managers, we 'touch up' the vines, but ninety percent of quality is from the site and the weather that year. Monte Rosso just happens to be a truly magnificent site!"

As I drove back down the hill, steering my SUV through the various blocks of Monte Rosso, I marveled again at the visionary spirit that created such a magical vineyard high on top of the Mayacamas Mountains so many years ago. Just before I reached the end of Monte Rosso, I stopped my vehicle next to a grouping of old *head-pruned* vines on my right. Though it was cold, the sun shone brilliantly on the black curling limbs, and the green grass shimmered with tiny drops of dew sparkling in a rainbow of colors around the massive roots of the vines. Yes, magical was a good word for this place.

Monte Rosso Signature Wine—Gnarly Vine Zinfandel

A few weeks later, I had the opportunity to meet with Mike Martini to discuss a signature wine of Monte Rosso. Though the Louis M. Martini Winery in Napa makes several wines from Monte Rosso fruit, the one that is most representative of the vineyard is the *Gnarly Vine Zinfandel* made from the 120-year-old vines in the Gnarly Vine Block.

As soon as I entered the tasting room of the Louis M. Martini winery, a very tall and impressive man with a full head of silver-white hair strode forward to greet me. He was wearing khaki pants, a blue shirt, and a dark-green down vest with the logo Louis M. Martini Winery on the upper right-hand side.

"You must be Mike Martini," I said.

A huge grin lighted up his face, and he grasped my hand in a hearty handshake.

"Yes, I am," he said. "Why don't we have a glass of Gnarly Wine while we are talking about how it was made." He gestured me to a chair in the tasting room, and he poured two glasses of a deep red-purple wine that was so dark and opaque, no light showed through it.

We both took a sip and savored the wine, not speaking for a few minutes so we could appreciate the bouquet, flavor, texture, and finish. The nose was massive, with ripe, opulent, dark berries and spice. On the palate the wine was rich with velvety tannins, intense concentration, and flavors of ripe plum, pepper, and a distinctive savory note that added to its complexity. The finish was very long, satisfying, and warm due to an alcohol content of over 16 percent. Despite this, the acidity was high and quite refreshing. This was no wimpy Zinfandel.

"So you want to know how I made this wine," Mike interrupted my thoughts.

"Yes, please!"

"Well, a lot of it has to do with Monte Rosso," he responded. "I love that vineyard. I remember, as a boy, riding on an old truck up the twisting road to Monte Rosso. We used to play in the vines as children. Have you ever seen that movie with Rock Hudson, *This Earth is Mine*? It was filmed on that road." Mike took a sip of his wine and looked off in the distance as if he was remembering those long-ago days.

"Which parts of Monte Rosso are your favorites?" I asked.

"I have several favorite blocks in the vineyard. For me, the Thieves was always rather spiritual. Not sure why—perhaps because there may be Indians buried there."

"Have you ever seen a rattlesnake in the vineyard?"

"We used to find close to sixty rattlesnakes a year in Monte Rosso, but they were always quite small—babies. They hide in the leaves of the vine, and we cautioned workers to never put their hands into the leaves without checking for snakes first. Luckily no one was ever hurt, but we did have a dog or two bit by a rattler. The old-timers always said that the canyon below the vineyard was a nesting place for rattlers, and there was a 'snake ball' of babies that climbed out of the canyon each year and went into the vineyard."

I shuddered at the image, and Mike continued. "I once saw a very large kingsnake in that vineyard. You know, they eat rattlesnakes," he said with a smile.

"What are some of the special aspects of Monte Rosso?" I asked, trying to move the subject off snakes to something more pleasant.

"One of the great things about Monte Rosso is it always delivers a nice high acid. Some people say it is the soil, but I believe it is the aspect. The vineyard is facing southwest, and the wind sweeps in each day and keeps the temperature within the canopy cool."

"Do you find differences in the Zinfandel from separate blocks?"

"Yes." He paused to take another sip of his wine. "I find that the grapes grown on top of the mountain plateau taste radically different from the terraced hillside grapes. The plateau grapes are more delicate and fragrant, whereas the ones on the slope are denser, with bigger tannins."

"So when you make Gnarly Vine, how do you decide when to pick the grapes?"

"I always wait until the berries are about two percent shriveled," he responded. "You know, Zinfandel always is uneven ripening. I don't go by numbers. Instead, I use sight and taste. My dad always said you should squeeze the berries to see if they run red. In the end, I always pick around twenty-six to twenty-seven degrees Brix, but by the time they are in the tank, they usually measure twenty-nine to thirty degrees. Would you like to see where we receive the grapes?"

I nodded quickly, and he motioned for me to follow him out the side door of the tasting room toward the winery in the back. We stopped at the grape-receiving area, where a variety of different stainless-steel tanks, wooden barrels, and presses were arranged near an overhead sign that read Cellar 254.

"This is the microwinery I had built," said Mike proudly. "It is where we make all of our small lots from special vineyard blocks. When the Gnarly Vine grapes arrive, we have a sorting table set up here with four people," he gestured to a spot on the cement where the sorting table would be. "I always ask them to remove the raisined grapes. I don't like raisins."

Mike walked me through the winery, pointing out all of the special equipment used to make the Gnarly Vine Zinfandel.

"Once the grapes are sorted, we put them through a special destemmer machine, a Laplenec, which has fingers that gently comb through the clusters. We end up with seventy-five percent whole berries. I then put them into small, stainless-steel containers, keeping all of the blocks separate. They go through a *cold soak* (where the must is kept cold to extract more color and flavor) for three to five days with dry ice at around fifty to fifty-five degrees Fahrenheit. Then we submerge the *cap* (top of the grapes) with a screen and raise the temperature so that there is some carbonic maceration—about forty percent of the batch— which gives the wine a nice floral aroma."

I was surprised to hear about the *carbonic maceration*, which is the method of fermentation that takes place inside the individual grapes in a carbon-dioxide environment *before* they are crushed. With traditional alcoholic fermentation, the grapes are usually partially crushed first. Carbonic maceration results in a fruitier wine with softer tannins.

"Why do you use partial carbonic maceration?" I asked.

"My goal is to create a Zinfandel with a fresh, brambly taste," he explained. "I don't want any jammy zins. The temperature is usually around seventy-five, but toward the end, we increase it to around ninety to ninety-five degrees Fahrenheit. Generally, I like to use a commercial yeast that they use in Barolo that can withstand the high temperatures. The reason I like to use high temperatures is to 'get the extract.' I want the texture and velvety tannins this technique provides. I actually do this with all of my wines, including Cabernets, Merlots, and Pinot Noirs."

Mike continued to explain how they punch down the fermenting Zinfandel twice a day with a mechanical plunger until it is necessary to do it only once a day. The complete maceration lasts around twenty-eight days, including a period of *extended maceration* (allowing skins to stay on after alcoholic fermentation is complete) of three to five days with dry ice. They use a basket press to gently press the skin off the wine and then transfer the wine to 70 percent new French oak barrels and 30 percent new US oak barrels for eighteen months.

The wine undergoes *ML* in the barrel. ML is short for *malolactic fermentation*, a secondary fermentation in which the malic acid in the wine is transformed into a softer lactic acid by a special type of bacteria. The majority of red wines and some white wines, such as Chardonnay, go through ML, which results in a softer feeling on the palate and can add a creamy, buttery character.

The wine is usually *racked* twice a year, which means transferring it to a new barrel so the sediment is left behind, and the wine is clearer. The sediment is often referred to as the *lees*. The wine is also topped in the barrel once a month with new wine so very little oxygen can harm it.

"We do not *fine* the wine," reports Mike, "but we will apply a light *filter*." *Fining* refers to the process of adding a fining agent such as egg white or bentonite clay to the wine, which binds to any particles and pulls them to the bottom of the barrel. The wine can then be racked to a new barrel and is clearer in color. *Filtering* involves running the wine through a light filter to remove any tiny particles.

Blending usually takes place during the first racking because they move the wine into stainless-steel tanks.

"We generally add about seven percent Petite Syrah," said Mike, "but it depends on the vintage. Since the Monte Rosso Zinfandel vineyards are old field blends, we are also picking up some of the mixed blacks that help provide this dark color and great complexity. I'm sure there is some Folle Blanche, Grand Noir, and Alicante Bouchete in the blend. It is the magic of the block."

After bottling, the wine usually ages for another year before release. "We only make about eight hundred cases of this Zinfandel each year," said Mike. "You know, some people say that Zinfandel will not age like Cabernet, but we've done tastings of old 1945 vintages of both, and you can't tell them apart when they get that old. It is fascinating."

To Kalon Vineyard
Oakville AVA, Napa Valley

Most people arrive at To Kalon Vineyard by driving along the famous Napa Highway 29, but I decided to brave the winding and stomach-churning Oakville Grade that switches back and forth through the Mayacamas Mountains between Sonoma and Napa Valleys. My purpose was twofold: one, it would be fifteen minutes shorter when driving from Santa Rosa, and two, it would afford me a breathtaking view of Napa Valley as I descended the slopes.

To Kalon Vineyard stretches from the trees at the edge of Robert Mondavi Winery on the north to south on Dwyer Lane. On the west it cuddles up against the Mayacamas foothills, and on the east it borders Highway 29. This means that the Oakville Grade cuts right across the middle of To Kalon Vineyard.

Naturally it was only a few minutes into the climb over the grade that I encountered a large, slow-moving dump truck that took me five patient minutes to pass on the steep and twisting, narrow two-lane road. However, once past the truck, I encountered no other cars on the bright and sunny but cold December morning, with a temperature hovering around 48°F. The drive was worth it, because as the Napa Valley came into view—stretched out in a multi-colored tapestry of yellow-, gold-, and red-leafed vineyards with tall mountains embracing its precious bounty—I sighed in appreciation, as I do every time I see that picture-postcard view. It is truly breathtaking, and To Kalon Vineyard makes up a significant portion of the splendor.

To Kalon is actually owned by three different entities, but the largest acreage is under the ownership of Robert Mondavi Winery. Therefore, I decided to begin my exploration by scheduling an appointment with Matt Ashby, the Director of Vineyard Operations for Mondavi To Kalon.

Portion of To Kalon Vineyard at Robert Mondavi Winery

We had agreed to meet under the famous winery arch at ten fifteen, and as I drove up, I saw a tall, brown-haired man in a tan jacket and jeans. He strode forward to greet me, and I could see that he had the quick, athletic movements of someone who spends much of his time outdoors. I knew from reading his bio that Matt had a degree in Agricultural and Environmental Sciences from UC Davis and had been working at Mondavi for over ten years. After a quick handshake and introduction, he ushered me toward a large pickup truck parked nearby.

"Let's go see the vineyard," he said with a smile, "and on the way I'll tell you a bit about the history of this place."

History of To Kalon

To Kalon was the brainchild of Hamilton (H. W.) Crabb, who left his native Ohio and traveled to Napa Valley with the intention of buying land to farm. When he arrived in 1865, Napa Valley was still predominately covered with long grasses, cattle ranches, and wheat fields. Within a few years, Crabb was able to buy land near the small village of Oakville, on which he planted orange and chestnut trees as well as table grapes. As the orchards and vineyards thrived, he decided to expand and eventually acquired over five hundred acres. In 1868 he started experimenting with wine grapes, and within a decade he had planted over two hundred different varietals on his property. Inspired by the success of the wine grapes, he decided to name the vineyard "To Kalon," which means "the highest beauty" in the Greek language.

By 1872, the vineyards of To Kalon were doing so well that Crabb established a winery and created a wine brand called To Kalon. Eventually, he ramped up production to over eight hundred thousand gallons (approximately thirty-three thousand cases), selling his wine across the United States and even France. His success placed him firmly among other famous early Napa Valley winemakers, including Krug, Beringer, and Schram. By the mid-1870s, Crabb was considered an authority on winegrowing. He was invited to write a chapter in a book in which he praised California as "the Vineland of the world" and an ideal place to make wine due to the warm, sunny climate.

In the early 1890s, *Phylloxera*, a microscopic insect that kills grapevines by eating the roots and leaves, invaded Napa Valley. To Kalon, like many other vineyards in the area, was devastated. Crabb worked to reestablish the vineyard, but when he died in 1899, it was sold at auction to the Churchill family. They continued to farm the vineyard and operate the winery until Prohibition.

Over the ensuing years, the property was bought and sold by several other owners, until one day in the 1960s, Robert Mondavi arrived. In his book, *Harvest of Joy*, Robert describes his first experience with the vineyard:

> *"Walking through To Kalon, admiring its contours and vines, smelling the richness of the soil, I knew this was a special place...As I walked, I felt a powerful, almost inexpressible connection to this land. And some root intuition inside me seemed to say, 'Yes, this is the place'" (p. 63).*

To Kalon Vineyard with Balloon

So Robert Mondavi purchased twelve acres of To Kalon Vineyard in 1966 and built his famous Spanish-style winery with the curving adobe arches, bell tower, and welcoming statue and fountain. Throughout the years, he continued to purchase more To Kalon acreage, so that by 2013, the winery owned 435 acres of the famous vineyard.

At the same time, sections of To Kalon were also purchased by others. Opus One, a joint venture formed by Mondavi and Mouton-Rothschild, owns 120 acres. Another famous portion of To Kalon is owned by Beckstoffer Vineyards, who purchased 89 acres from Beaulieu in 1993, and renamed it Beckstoffer To Kalon.

Therefore, today To Kalon is 654 acres and is one of the largest and most historic vineyards in Napa Valley. It is recognized as producing some of the highest-scoring California wines, winning multiple awards. Grape buyers representing other wineries compete with one another for the opportunity to be allowed to purchase To Kalon fruit.

Touring To Kalon Vineyard

Matt maneuvered the large pickup truck down a dirt road around the back of the winery.

"That large row of pine trees to the north of the complex is the property line for To Kalon," he said. "In total, the Mondavi portion of To Kalon currently encompasses 435 acres that is divided into 87 five-acre blocks."

We drove past several different sections, and I was surprised at how diverse they looked. Some vines were on very tight spacing with low trellising, whereas others were organized with wide spacing and trained on a variety of systems. There were also a few blocks where the vines had been ripped out and placed in large piles for burning.

"We believe in research and development here," explained Matt, "and that is why you will see so many diverse trellis systems and different spacing. We are currently replanting 130 acres."

Matt stopped the truck in the middle of a narrow dirt road at the back of the vineyard, near the foothills of the Mayacamas. We climbed out, and he pointed toward Highway 29 in the distance, where we could hear cars racing by.

"I want you to see how much the land slopes here," he said. "From where we stand to Highway 29, the elevation drops close to thirty feet. You are standing in the middle of an alluvial fan."

I looked down at the ground on which we stood and saw small pieces of gravel on top of the brown earth. The gravel was a variety of sizes, shapes, and colors, ranging from small gray stones, to dark black, brown, and tan pebbles.

"As you probably know," Matt continued, "an alluvial fan is formed when soil and gravel from the mountains flow down the hillsides and are deposited in the lower hills and valley floor in a fan shape. Here in the Napa Valley, this has resulted in a variety of soil structures that are ideal for wine grapes."

I was excited to realize that I was actually standing amid such a famous soil structure as an alluvial fan. Napa Valley's alluvial fans are famous in global wine circles, because it is known that the deep mixture of gravel and rocks causes the grapevines to push their roots far down in order to find nutrients and water. This causes just the right amount of stress to create high-quality wine grapes. Combined with the ideal Mediterranean climate of Napa Valley, this happy situation explains the Napa Valley Vintners Association saying of "To a wine grape, it's Eden."

Matt motioned for me to follow him into a nearby block of tall Cabernet Sauvignon vines with huge, thick trunks and stunning winter leaves in multiple hues of red, orange, yellow, and maroon. The colors against the bright blue sky were so vivid, I felt I was looking at a beautiful oil painting.

"These are magnificent," I exclaimed in awe.

"Yes, they are beautiful," said Matt, "but as you probably know, the red leaves mean they have a virus. However, they still produce excellent fruit, and even though the yields are lower, we plan to keep them awhile."

"I hope so," I said and wondered again about the grape leaf roll virus that impacted so many old vines. Though striking to see in the red leaves it produced around harvest time, it also slowed sugar production and reduced yields. Some vineyard managers replaced the vines, whereas others chose to appreciate the lower Brix (sugar content) they produced, arguing it was similar to the sugar levels they used to harvest in the old days.

Matt put out his hand and placed it on the large wooden stake upon which the vine had been trained. "As you can see," he said, "these are head-pruned vines, and they've been with us for many years."

Winter Cabernet Sauvignon Vines in To Kalon

As he stood next to the old Cabernet Sauvignon, I realized it was at least five feet tall with a trunk thicker than a man's arm. I could hear the pride in Matt's voice and knew that being the guardian of these vines was important to him.

"Come on. I'll show you something quite different now," he said and gestured for me to follow him down the dirt road and then across into a different section of the vineyard.

When we arrived it was almost as if we had changed continents. All of a sudden, we were standing in a very tightly spaced block with short vines about three feet tall. I had a sudden flashback to my visits to vineyards in Bordeaux that use similar spacing and trellis systems. Even the soil, which was covered with small bits of pebbles and rocks, was very similar to the gravelly vineyards of the Left Bank.

"So we're in France now?" I asked.

Matt laughed. "Sort of. This is one of our high-density blocks," he said. "The spacing is four by four feet, and the vines are trained on a low, double-Guyot trellis system."

"Why is this here?"

"At Robert Mondavi, we believe in experimentation and innovation," Matt said proudly. "Also, when we started Opus One, our joint venture with Château Mouton Rothschild, part of To Kalon went to Opus One, and they planted small-density blocks. We thought we would do the same. The philosophy is that because of the tight spacing, the vines are forced to compete with one another for nutrients, which usually results in lower production and smaller berries. The higher skin-to-pulp ratio usually means more flavor, more tannins, and darker color in the wine."

"So is it working?"

"Seems to be," said Matt, "but California is different from Bordeaux. We've been doing experiments for years with different spacing, trellis systems, rootstocks, and clones. With spacing, we've discovered that six by four appears to be the most efficient spacing with good quality."

"Interesting."

Matt turned and pointed across the road to another block. "That is one of our most famous blocks. It is called the I-Block, and it is composed of head-pruned Sauvignon Blanc vines planted in 1945."

I turned my head to look and was mesmerized. Here was a block of old vines that took my breath away. It seemed as if I was looking at a field of black wrought-iron candelabras turned upside down and placed in rows.

As we walked closer, I could see that the wide rows were filled with tall green winter grass, early yellow mustard, and green barley shoots and legumes. The thick, gnarled trunks rose toward a blue sky, while golden leaves still adorned some of the shoots.

"How beautiful."

Matt smiled. "Many people say that when they see this block. It is one of our most cherished blocks and was here when Robert Mondavi bought To Kalon. The grapes produced here go into the I-Block Fumé Blanc."

I thought of the many times I had tasted that delicious Sauvignon Blanc wine with notes of grapefruit, pineapple, and a cleansing acidity. Here was the source. I reached out to touch the tip of one of these ancient beauties and immediately felt a sense of peace and awe. The words that Robert Mondavi wrote in his book, *Harvest of Joy*, came flickering back into my mind:

> *"It seemed to radiate a sense of calm and harmony, of peace and serenity (p. 63)."*

"We have another block of old Sauvignon Blanc planted in 1954," said Matt with obvious pride lighting up his face. "It is also head-trained and dry-farmed on 11.6 acres east of the UC Davis Experimental Station." He gestured toward the east of the property. "It has been renamed 'Robert's Block,' in honor of Robert Mondavi."

I-Block of Head-Pruned Sauvignon Blanc

Back in the truck, we headed southwest, crossing several other blocks that Matt said were owned by Opus One. I also knew that the eighty-nine acres of To Kalon that Beckstoffer owned were nearby, bordering part of Highway 29. Upon acquiring the property, the Beckstoffers had made many positive improvements, and now the grapes they produced fetched some of the highest prices of any vineyard in Napa Valley.

Eventually we came to a stream bordered by large leafy trees and green grassy banks. Matt drove the truck onto a small bridge and idled the engine, so we could look down at the sparkling water that bounced over rocks and twirled in small pools. It was an inviting and peaceful place, and I could imagine reading a book near the stream or having a picnic there.

"This is called Doak Creek," he said. "It was revamped as part of our environmental efforts. The water is now very clean and is filled with native fish part of the year. We've replanted many of the original California grasses and shrubs and have added raptor roosts to encourage hawks and other birds of prey to return."

As we continued our drive through the vineyards, I became more and more amazed at the vast size of To Kalon and how diverse the blocks were. Dark winter vines spread out across the valley on various types of trellis systems, different spacing, and changing soil colors and textures.

Eventually we reached the end of one block, and I realized we had come to Oakville Grade, the same road I had driven across a couple of hours earlier. Matt turned right onto the road and drove west toward the Mayacamas Mountains. Vineyards spread out on both sides of the road, and I realized I was still deep in the heart of To Kalon.

Matt parked the truck on the side of the road and pointed at the vineyard on the right-hand side. "This is the Monastery Block," he said. "It was named for the old monastery up there on the hill."

I squinted my eyes and could just make out the white tower of the monastery, surrounded by a grove of thick trees. It stood on the foothills of the Mayacamas overlooking this section of To Kalon Vineyard.

"What a perfect name for this block," I said.

"Yes," agreed Matt, "and this block produces some of our most intense Cabernet Sauvignon. As you can see, it is on four by four spacing as well."

I glanced out at the tightly packed vines and wondered if their position near the old Carmelite monastery gave them any special blessings. The monastery had been established in 1955 on the grounds of the Doak mansion, which was built in 1917. Surely, with the many prayers said within those holy grounds over the years, some of the positive thoughts must have drifted over the vineyard.

Next we crossed the road and visited several other blocks of vines near an old house, and then Matt took me to see the irrigation pond.

"This is huge," I exclaimed, as we stood next to a very large pond of rippling, dark-blue water.

"Yes, it is a nine-acre irrigation pond," reported Matt. "It has two filters, and we clean and recycle rainwater here."

"Impressive."

"It is part of our sustainable farming practices to recycle water," he continued. "In addition, because To Kalon is in such a perfect location, we found that most years we don't need to irrigate the vines until mid-July."

Vineyard Specifics for To Kalon

As we headed slowly back toward the Robert Mondavi tasting room, Matt provided many details on the specific aspects of the vineyard. The primary grape *varietals* planted in To Kalon are four of the major red Bordeaux grapes: Cabernet Sauvignon, Petit Verdot, Cabernet Franc, and Merlot, with a small amount of Malbec. Robert Mondavi Winery still maintains about 20 percent of its acreage in the two white varietals of Sauvignon Blanc and Sémillon.

In terms of *soil* structure, since To Kalon is part of one of the famous alluvial fans in Napa Valley, it actually has seven types of soil: Bale Loam, Bale Clay Loam, Clear Lake Clay, Coombs Gravelly Loam, Yolo Loam, Haire Loam, and Perkins Gravelly Loam.

The *elevation* is 110 feet where the vineyard borders Highway 29 but rises up to 140 feet where it ends at the foothills of the Mayacamas. Summer *temperatures* average 84°F for a high and 51°F as a low. *Rainfall* is usually around 44 inches per year.

Vineyard Specifics for To Kalon

Total Vineyard Acres	435 (plus Opus One: 120; Beckstoffer To Kalon: 89)
Varietals	Cabernet Sauvignon, Petit Verdot, Cabernet Franc, Merlot, Malbec, Sauvignon Blanc, and Sémillon
Soil	Bale Loam, Bale Clay Loam, Clear Lake Clay, Coombs Gravelly Loam, Yolo Loam, Haire Loam, and Perkins Gravelly Loam
Elevation	110 to 140 feet above sea level
Average Temperature	Summer temperatures average 84°F for a high and 51°F as a low. Rainfall is usually around 44 inches per year.
Rootstocks	St. George, 03916, 101-14, 110R
Clones	Sauvignon Blanc: clone 1 Cabernet Sauvignon: 4, 6, 7, 15, 169, 337, and the Mondavi Heritage Clone
Sun Exposure	Northeast by southwest line
Spacing	10 x 8 for older vines; 4 x 4 for high density, 6 x 4 for newer plantings
Trellis Systems	Bush vine, VSP, modified California sprawl

Rootstocks and clones vary throughout the blocks. According to Matt, "We do not want to have all the same rootstock because we want to spread the risk and not get caught with a rootstock that isn't resistant to certain diseases as we experienced with *Phylloxera* and AxR1 in the 1990s. Further, the soil is different in certain blocks, requiring distinctive rootstock." Therefore, the vineyard includes some of the original St.

George rootstock, such as that used with the I-Block of Sauvignon Blanc on clone 1.

"In addition," Matt reported, "To Kalon has 03916 rootstock, which is resistant to nematodes; 101-14, which I believe is well-balanced and not too temperamental; and 110R, which is drought resistant."

Clones vary by varietal, but some of the Cabernet Sauvignon clones include the traditional ones of 4, 6, 7, and the Mondavi Heritage Clone, as well as a few of the new, certified virus-free clones: 15, 169, and 337.

In terms of *sun exposure*, the majority of the Mondavi blocks are planted on a northeast by southwest line to avoid the worst of the summer heat that could cause sunburn to the grapes.

Spacing and trellis systems vary based on varietal and age of the block. Older blocks are on wider spacing such as ten by eight feet, whereas high-density blocks are four by four feet. Trellis systems also vary; the most common, VSP (vertical shoot positioning), uses the spur-pruned double cordon; however, older vines are head-pruned.

"As mentioned, we are moving toward six by four as the most effective spacing," commented Matt, "and a modified VSP system with crossarms, which allow for flexibility in moving shoots between the wires to create a modern California sprawl. This helps us position the canopy to allow the leaves and fruit to have the ideal sun and air exposure."

Farming Practices—Certified Sustainable and Fish Friendly

To Kalon Vineyard is *certified* as a Fish-Friendly operation and by the California Sustainable Winegrowing Alliance.

"I am really proud of all of the advances we have made in environmental protection here," says Matt. "Not only have we revamped Doak Creek to bring back the salmon, but we have added many native plants to attract beneficial insects such as ladybugs, which are great for the vines. In addition to the raptor perches I mentioned earlier, we have put in owl boxes that help to reduce gophers, voles, and bird damage to the vineyards. Because of this, we do not have to use bird netting or traps."

For *fertilization*, the major method is to plant cover crops of barley, legumes, and mustard that are plowed back into the soil to provide nutrients for the vines. Mondavi also makes its own compost that is used as fertilizer and will use a combination of organic and nonorganic commercial fertilizer, if needed. Petiole analysis is conducted every spring to determine if additional nutrients are needed.

Farming Practices at To Kalon

Certifications	Fish-Friendly Farming, California Sustainable Winegrowing
Fertilization	Cover crop, organic compost, and other organic and nonorganic fertilizers as needed
Weed Control	Mowing, dormant herbicide sprays
Canopy Management	80 percent suckered and pruned by hand, the rest mechanical
Disease Control	Organic and traditional fungicides to prevent powdery mildew and *Botrytis*
Pest Control	No major issues, using owl boxes; new rootstocks to control nematodes and mealy bugs
Irrigation	Drip- and dry-farmed on older vines; have five wells
Technology	Weather stations, wind turbines, neutron probes
Harvest Measurements	Winemaker determines by taste and lab measurements

In terms of *canopy management,* more than 80 percent of the vineyard is suckered and thinned by hand, but 20 percent of the blocks can be prepruned, hedged, deleafed, and even harvested mechanically. *Weeds* are controlled with dormant herbicide sprays such as Roundup.

Pest and disease control do not appear to be much of an issue at To

74

Kalon. "Due to our sustainable winegrowing efforts," explained Matt, "we have very little problem with traditional pests." Mondavi Winery protects the vines from powdery mildew and *Botrytis* with a combination of organic and traditional fungicides. "Our main challenges," continued Matt, "are fan leaf disease caused by nematodes and leaf roll, which can be transmitted by mealy bugs. This is why we are replanting blocks on rootstocks that are more resistant to these issues, as well as using some of the new virus-free clones."

To Kalon is equipped with an impressive *irrigation system and technology sensors*. They have drip irrigation in all blocks with the exception of the older head-pruned blocks, which are dry-farmed. In addition to the large irrigation pond, they have dug five water wells and have installed nine neutron probes to check for moisture. These probes measure nine feet into the soil and provide valuable data throughout the growing season. They conduct leaf-water-potential analyses on a weekly basis.

"Despite all of the technology," said Matt, "for the most part, we use our five senses to determine if additional irrigation is needed. For example, if the wind is picking up and my lips feel chapped, then I am concerned about lack of water, but if the fog is rolling in, then I know the vines are getting some moisture." To Kalon also has multiple weather stations and wind turbines to protect from frost.

Measurements for harvest are determined jointly by the vineyard and winemaking teams. "Genevieve, our winemaker," said Matt," spends much time in the vineyard with us and is here almost daily, starting in August. She tastes the grapes and has the lab run measurements. Obviously we are looking to harvest the grapes with ideal fruit and tannin maturity, as well as acid and pH levels, but the timing of those characteristics changes with each vintage."

Economics of the Vineyard

When I asked Matt about vineyard costs, he was able to provide the information with quick and impressive accuracy. "The average annual cost for *farming costs*, including labor, at To Kalon," he said, "is approximately nine thousand dollars per acre or around two thousand five hundred dollars to three thousand dollars per ton. With one hundred and thirty acres currently under development, the average cost to install a new block is twenty-five thousand dollars per acre."

All vineyard labor is outsourced to a labor contractor with average rates ranging from $10 to $16.50 per hour depending on the type of work. For harvest, workers are paid at a piece rate ranging from $95 to $150 per ton.

"A talented labor team," reported Matt with pride, "can easily pick ten tons per day."

Average yield is dependent on the block composition, with older head-pruned blocks producing less. However, overall, To Kalon's average rate is 2.9 – 3 tons per acre. When all 435 acres are in production, this calculates to approximately 1305 tons.

Economic Viability of To Kalon Vineyard

Average Yield	2.9 – 3 tons per acre
Total Average Tons Per Year	1305
Costs Per Acre	$9,000
Revenues	Not available
Economic Health	High

Revenue from To Kalon grapes is difficult to calculate exactly, since most of the fruit goes into Robert Mondavi Wines, with a small amount sold to Opus One. The Robert Mondavi Winery in Napa Valley produces approximately two hundred thousand cases annually within four tiers: Napa Valley (retail $20–$28), District (retail $40–$85), Reserve (retail $40–$150) and Spotlight (retail $40–$250).

Grapes for the first two tiers are purchased from neighbor vineyards, but To Kalon Sauvignon Blanc is used exclusively in all four tiers. To Kalon Cabernet Sauvignon and other Bordeaux varietals are used in all tiers except for Napa Valley. Only the most exceptional blocks in To Kalon are used in the Reserve and Spotlight tiers, depending on quality.

According to Matt, To Kalon Vineyard has a high rate of *economic viability* and is doing quite well in terms of return on investment, especially at the top tiers that produce wine over $100 per bottle.

The Soul of the Vineyard—"Beautiful"

As we neared the conclusion of our tour of To Kalon, I asked Matt several questions to tap into his feelings about working there, including one word he would use to describe the vineyard.

"If I had to describe To Kalon in one word," he said, "it would have to be *beautiful*. It is so unique and amazing. For example," he continued, "the I-Block of Sauvignon Blanc takes you back in history to the 1940s. I gaze at it and think, 'this is living history!' Sometimes I have to pinch myself to believe I'm working here. I am so lucky."

When asked about the best part of working the vineyard, a large smile spread across Matt's face. "That's easy," he answered. "It is tasting the wine. When I sip a glass of wine that was made from To Kalon, I feel wonderful—because I know I had a role in helping to grow the grapes that go into such amazing wines."

In terms of challenges, Matt describes the angst of trying to unravel the mystery of the different blocks. "We are always trying to understand what we can do to bring all of the blocks to reserve quality, but in the end, Mother Nature rules." Matt paused to look out across the vast vineyard. "For example, this one block has poor drainage and a different soil mix from that one over there," he gestured toward a block closer to the foothills. "Though we have put in a drain, those vines still don't produce enough quality to go into our top program. I am continually asking myself, 'what's happening here' and 'what can I do to help improve quality.'"

Regarding key learnings from working the vineyard, Matt responded readily. "I have learned to respect the soil and to try to be a steward of the land. Working in this historic vineyard, I recognize that it is a special jewel that needs to be protected. My role is to work with the soil, the vines, and nature to help the vineyard stay healthy."

As the famous bell tower of the Robert Mondavi Winery came into view above the vast sea of vines, I knew our tour was ending. Therefore I tried to tap into my impressions of the famous and historic To Kalon Vineyard. Massive and diverse were the first responses that came to me. To Kalon seemed like a huge tapestry or quilt with many colors and patterns—so many blocks, so much experimentation, and so much continual learning taking place all of the time.

On the other hand, there was a sense of peace and harmony in the vineyard—especially in the blocks of old vines. It was more like a vast community of living creatures that coexisted and learned from one another: new blocks, old blocks, and communities living with different architectures of trellis and spacing and various clonal backgrounds. It was a diverse tapestry, and I had the feeling that To Kalon was actually enjoying all of the experimentation. There was a sense of creativity and fun here—with Mother Nature, science, and art all intersecting in the colorful quilt that was To Kalon.

Signature Wine: To Kalon Reserve Cabernet Sauvignon

Though Robert Mondavi Winery has produced many exceptional wines throughout the years using grapes from To Kalon, it is the Reserve Cabernet Sauvignon for which they are most known and have received the largest number of rave reviews and rating. Therefore, I focused on this wine and the 2006 vintage that received 96 points from *Wine Spectator* and sold for $135 per bottle upon release.

The head winemaker at Robert Mondavi Winery in Napa Valley is Genevieve Janssens. I have met her on several occasions and have always been impressed with her confident style. With her flashing, dark eyes; wavy, black hair; and French accent, she is passionate about the wines she makes from To Kalon grapes.

"It is a privilege to work in To Kalon," said Genevieve. "I have been here since 1978, and it is my driving reason—raison d'être—for working at Mondavi. I am always discovering new things about the vineyard. For example, through the years I've discovered that To Kalon is reacting positively to weather events. To me, To Kalon is the superstar of Napa Valley. It is a First Growth vineyard."

In crafting the 2006 Reserve Cabernet Sauvignon, Genevieve describes her time in To Kalon Vineyard. "I walk the vineyard every day starting in August. I am looking at the leaves, the clusters, and the soil— and of course, tasting, tasting, tasting. I don't look at the numbers myself, though of course they are measured in the lab. I am more interested in the visual and the sensual. It is the relationship between the meat and the skin of the grapes. I also work very closely with the vineyard management team. We believe in teamwork."

The weather in 2006 was cooler than normal in Napa Valley, but there were four days of heat spikes over 100°F in July. Because of this, harvest of the Cabernet Sauvignon started later, on October 12. According to Genevieve, "The grapes for the 2006 came from an older block of dry-farmed, head-pruned Cabernet Sauvignon on St. George rootstock." In the end, the average Brix was 26° with 3.52 pH and .71% initial acid. In addition, Cabernet Franc grapes were also harvested for this wine, with some of the grapes coming from a nearby vineyard in Oakville.

Grapes were hand-harvested and hand-sorted in the cellar and then destemmed and lightly crushed. "I like to cold soak for 7 to 10 days in 210-hectoliter foudres (large barrels)," said Genevieve, "and keep them half-filled and protected with dry ice around 10°C (50°F). We then raise the temperature to about 15°C (60°F) and inoculate with commercial yeast. Alcoholic fermentation usually takes around 10 days at 30°C (86°F). We don't allow it to go higher than 32°C (86°F). When that is finished, we start a 10- to 15-day extended maceration and keep the wine covered, so no oxygen can get in."

In the end, the 2006 vintage had a total maceration time of thirty-seven days before being pressed off gently in a traditional basket press. *Maceration* refers to the total time the grapes are soaked with their skins, including premaceration in a cold soak before fermentation starts, as well as extended maceration after alcoholic fermentation has stopped. A longer maceration time usually results in a wine that has deeper color and more intense flavors.

Genevieve then had the wine transferred to 100 percent new French oak barrels where it underwent malolactic fermentation and aged for twenty-four months. The barrels had a *medium toast*, meaning the wood inside the barrels was lightly charred, allowing the wine to take on a moderate amount of toasty oak flavors. French barrels are generally more delicate in flavor than US or other types of oak. Genevieve reported that the French oak "imparts more of a honey note with a touch of allspice." US oak, on the other hand, is often reported to give a broader, sweeter note of coconut, crème brûlée, and even dill at times.

"We generally rack around five times," continued Genevieve, "and generally do not fine the wine because the tannins in Napa Valley are so silky and ripe. I believe in Bordeaux it is necessary to fine because some vintages have more green tannins."

The final blend, consisting of 92 percent Cabernet Sauvignon from To

Kalon and 8 percent Cabernet Franc, is a perfectly balanced wine with complex notes of ripe plum, tobacco, rosewood, and black olive. Even though the alcohol is a high 15.3 percent, Genevieve and her team have done an excellent job of balancing this with fine-grained tannins; deep, concentrated fruit; medium plus acid; and a very long finish.

As I sip the wine again, I cannot help but think of Genevieve's comment that To Kalon is like a First Growth vineyard in Bordeaux. Palate memories of tasting the five Grands Crus Classés wines of Latour, Lafite, Margaux, Haut-Brion, and Mouton Rothschild cross my mind. Yes, this magnificent To Kalon Cabernet Sauvignon does have the same concentrated depth, balance, and well-integrated massive tannins of those wines. But there is something extra—a rich, red, velvety fruit that speaks clearly of the sunny skies of Napa Valley and "the highest beauty" produced by To Kalon.

Early Spring

"Bud Break"

Seghesio Home Ranch Vineyard

Stag's Leap Vineyard

Hirsch Vineyard

Chapter Five

Seghesio Home Ranch Vineyard
Alexander Valley AVA, Sonoma County

The Canyon Road Exit appeared sooner than expected as I drove north on Highway 101 on a sunny afternoon in late March. My destination was Seghesio Home Ranch Vineyard, which is approximately halfway between Healdsburg and Cloverdale. Turning left to drive under the freeway I found Chianti Road and drove several more miles north until the gnarled black trunks of old Zinfandel vines appeared on my left. Tiny pale green leaves sprouted along the shaggy arms of the vines, and my heart soared, as it always does, when I see bud break on ancient vines. It is a perennial sign of hope and rebirth that the twisted dead-looking branches burst forth in a froth of green lace each spring.

As I turned the wheel to drive up the narrow dirt road, I noticed patches of white daisies, tall yellow mustard and delicate bell beans with pink flowers growing in the rows between the vines. Ahead was an old Victorian one-story house painted in a faded goldenrod yellow, and in the front yard stood two glossy green trees filled with oranges. Wrapping around the front and side of the house was a wide porch trimmed with a delicate white railing that matched the gingerbread trim dripping down in a wood-scalloped pattern from the eves. It was then that I realized the vineyard completely surrounded the house in a protective circle before marching off to climb the distant hills.

Suddenly it dawned on me how appropriate the name of Home Ranch Vineyard was to this plot of land. The house was the original homestead of the Seghesio Family who bought the land in 1895, and planted the vineyard to surround their home. There were several other buildings near the house, including a barn-like structure and what appeared to be an old train station waiting room with the words "Chianti Station" emblazoned on the front.

Looking around I saw a silver pick-up truck was parked near the barn, and as the door to the truck opened, a man in blue jeans, red checked shirt, and brown work boots climbed out. He approached me to shake my hand in hearty warm grip, and I noticed he wore tinted glasses that hid the color of his eyes, but the smile on his face surrounded by a white mustache and beard made me feel very welcome. Over his thick gray hair, he wore a faded olive green baseball cap that had a red logo with the words *Vineyard Industry Products*.

"Welcome," he said.

"Thanks. You must be Ned."

"No, I'm Jim. Ned will be here shortly."

"Sorry to get you two mixed up," I said feeling a little confused. I knew I was meeting with the father and son vineyard team for the Seghesio vineyards, both related to the family by marriage, but for some reason I had the names reversed.

"No problem," chuckled Jim. "Happens frequently. I'm Ned's father, married to Julie Seghesio." He turned to gesture at the buildings "This is the old family homestead, but no one lives here anymore since the winery was moved to Healdsburg."

I nodded, remembering my visits to the Seghesio tasting room in the charming wine-centered village of Healdsburg.

"But this is the famous Home Ranch Vineyard that won all of the awards for its Zinfandel, right?" I asked, looking around the property.

"Yes," said Jim, "and this building here is where the original winery was located." He pointed towards the barn-like structure I had noticed earlier. "I used to help out here when I was younger and we were in the bulk wine business. I assisted with bottling and cleaning out tanks sometimes, but I much prefer working in the vineyards. I like to feel the soil under my feet."

"That's good." I flashed him a smile, pleased to hear of his passion for the vines. "What is that building over there that says Chianti Station?"

"Oh," said Jim turning to look at the structure. "That's the old railroad station for Chianti."

"Chianti? But I thought this was a Zinfandel vineyard. Chianti is made with the Sangiovese grape."

Jim grinned. "Guess you don't know all of the history of this place then?"

History of Seghesio Home Ranch Vineyard

In 1886 Edoardo Seghesio left his home in the Piedmonte region of Italy and immigrated to the US. Eventually he arrived in Sonoma County where he took a job working at the Italian Swiss Colony Winery for seven years, located in the beautiful Alexander Valley.

The winery had been established by Andrea Sbarboro in 1881 as an agricultural coop to provide work for Italian and Swiss immigrants. It eventually became so large and famous that it attracted a large number of tourists in the early 1900s. As the winery grew, Andrea established two communities in the area. One he named Asti and the second one was called Chianti, after famous places in his homeland of Italy. When they started to ship wine via the railroad to the East Coast, train stations were erected in each community, and therefore Chianti Station was established.

It was across the road from this station that Edoardo Seghesio purchased 56 acres with a small house in 1895. He brought his bride of one year, Angela Vasconi, to live on the property and together they planted a Zinfandel vineyard. It is thought that Edoardo brought the vines from the Italian Swiss Colony, because they were growing Zinfandel at the time. The grape was very popular in California at the turn of the century because it could make a wine that was ready to drink in one year, rather than be aged for a long time as other varieties required.

Seghesio Family Homestead

By 1902 the Seghesios had constructed a winery on the property and began making wine. The following year they were ready to ship some of the wine by railroad to San Francisco from the little Chianti Station near their house.

"There are stories," reported Jim, "that Angela used to sit on the front porch of the house and watch people disembark from the train. If she saw it was a wine buyer, she would have the hired man go out to coop to kill and clean a chicken. Then she would cook it and get a nice meal ready for the buyer. In that way she helped to sell a lot of wine." Jim grinned widely.

As the years passed, the Seghesios purchased more land around their house and the Chianti train station. In 1910, Edoardo planted a Sangiovese field blend that included the traditional white blending grapes of Trebiano and Malvasia. Today this is the oldest planting of these varietals in the US, and some of the vines still exist.

"The Chianti Block," explained Jim, "is just there across the road where the old Chianti Station used to me." He pointed with his index finger at a vineyard on the other side of Highway 101, and I followed his direction. Narrowing my eyes I could just make out some old vines growing tall and twisted through the bright green grass.

"So they did grow Sangiovese here after all?" I asked.

"Yes, and still do. We have several blocks of sangio on this property, all grown from the original budwood."

"That's fabulous," I said. "But why is the Chianti train station here now instead of across the road near the railway tracks?"

"Because they had it brought over a number of years ago when this winery was still in operation," said Jim, taking off his hat and running a hand through his thick hair, before replacing the cap. "It's part of the family history."

"So what happened next?"

Six months before the beginning of Prohibition the Seghesio winery was doing so well that Edoardo and Angela decided to purchase the Italian Swiss Colony Winery, which had recently been put up for sale. When Prohibition hit, they took on other investors, hoping to ride out the storm, but eventually had to sell Italian Swiss Colony to fellow partners.

Edoardo passed away in 1934, one year after Prohibition ended, and Angela took over control of the Seghesio Winery. With the help of her sons, Pete and Art, she continued to operate the winery and purchase more vineyard land whenever possible. In 1949, they acquired the old Scatena Winery in Healdsburg, and expanded their production.

When Angela passed away in 1957, Pete and Art took over management of the winery, and continued the tradition of vineyard acquisition and expansion. By the 1960s, they produced wine from more than 50% of the red grapes in Sonoma County and became one of the largest bulk wine operations at 1,700,000 gallons.

Finally in 1983, the Seghesio's decided to bottle some higher quality grapes under their own name. Pete Jr. Seghesio, son of Pete, and Ted Seghesio, grandson of Art, both took leadership roles in shifting the focus from bulk wine to high quality handcrafted lots. In 1993, the family made a decision to only bottle grapes from their own estate vineyards, and reduced production to 30,000 cases. The decision paid off, as the reputation for quality wines grew. By the early 2000s they were achieving 90+ points in multiple wine connoisseur journals, and were awarded 95 points for the 2009 Alexander Valley Home Ranch Zinfandel in *Wine Spectator*.

Today all winery operations and the tasting room are located in the site of the old Scatana winery on Grove Street in Healdsburg. They still focus on producing 30,000 cases of high-end estate wine, but also bottle an additional 115,000 cases of Seghesio Sonoma County Zinfandel from a variety of vineyards around the county. In 2011, the Seghesio Winery was sold to the Crimsom Wine Group. However, family members still remain very involved, with Ted Seghesio as winemaker and Jim and Ned Neumiller in charge of vineyard operations.

Touring Home Ranch Vineyard

"So now that you know the history, let's go visit the vines," said Jim. He led me to the edge of the vineyard block behind the house and I looked with pleasure at a group of old head-pruned Zinfandel vines. They stood about three to four feet tall and were a collection of different shapes and sizes with twisted black trunks and limbs covered with the fuzz of newborn leaves. Interspersed every so often along the rows of vines was an obviously newer vine, as attested to by its slimmer size.

"So these are both old and new Zinfandel vines?" I asked.

Jim nodded. "Yes, the family has a policy to never take out a vine until it dies. This is because they believe in the high quality and power of old vine zin." He turned and looked directly at me. "But when one does die, we replace it with a new one crafted from the same budwood as the original."

"And what is the original rootstock and clone?"

"We think most of the rootstock is St. George, but we're not sure of the clone. Since it is so old, we call it 'Edoardo's Heritage Clone.'

"That's clever," I said, recognizing that after vines have been in a vineyard for many years they can gradually change and mutate into something unique. "How is it different from other Zinfandel?"

"Well," said Jim, "it has moderate length clusters." He held out his hands to illustrate a measurement of about eight inches in length. "It has relatively small berries and no shoulders."

"That sounds positive for zin," I said, "given that the varietal usually has long clusters with shoulders." This larger size and the tendency to ripen unevenly made Zinfandel a more complicated grape, at times, for winemaking. Therefore this unique clone of Edoardo's, with its smaller berries and clusters, most likely created more concentrated wines that contributed to higher quality.

Jim nodded. "We're not sure where this Zinfandel clone came from originally, but it has adapted to the land here, and produces beautiful wines."

"So how old are the vines?"

"Well as you know the original vineyard was planted in 1885, but most of the old vine Zinfandel you see here is from 1910 to 1940. Right now we are at about 50% old vines intermixed with the new. Tractor Blight got some of the vines."

"Tractor Blight?" I was puzzled, because I had read about many vineyard diseases but never this one.

Jim grimaced. "That's what we call it when a tractor accidently cuts or knocks over a vine. It doesn't happen as much anymore as in the old days, because we are more careful and use better equipment. Besides now we want to preserve the old vines."

"Well, I hope so," I muttered, thinking of the additional premium old vine Zinfandel contributed to growers and winemakers alike.

"Hey Dad."

We both turned at the voice behind us, and I saw a younger version of Jim approach. He wore the standard blue jeans, work boots and matching hat of his father, but instead of a red-checked shirt he wore a light grey T-shirt with a *Farm Bureau* emblem.

Old Vine Zinfandel at Seghesio Home Ranch

"You must be Ned," I said stretching out a hand in greeting, and noticing that he had hazel eyes and the same mustache and beard as his father but in a rich brown color.

"Yes I am," he smiled in greeting. "Has Dad told you about Rattlesnake Hill yet?"

"No." My eyes widened and I look quickly down at the ground to see if there were any snakes around.

"Not here," said Jim with a laugh. "Rattlesnake Hill is over there." He pointed in the distance to where a small hill rose rather steeply and was covered with terraces of vines on VSP trellising that were strung across the green hill like a row of Christmas lights. The hill was perfectly cone-shaped from where we were standing, and it reminded me of a miniature version of the Hill of Corton in Burgundy.

"That's part of Home Ranch Vineyard?" I asked surprised, thinking it was someone else's property.

"Yes," said Jim. "The whole vineyard is actually 190 acres and spreads up and around that hill, as well as over another hill further south and jumps across Highway 101 to encompass the old Chianti Block."

"Wow, I didn't realize it was that big."

"Yes, and that's just one of our vineyards," said Ned with obvious pride in his voice. "We have six altogether with around 350 acres to farm."

"Impressive," I said. "But what about rattlesnakes?"

"Well," chuckled Jim, stroking his beard with his thumb and forefinger, "we call it that because when we were putting in some new vines of Sangiovese we found a nest of rattlesnakes when we moved a large rock. Hundreds of snakes swirled around on the ground that day, and then raced down the other side of the hill and into the canyon."

"Ugh," I couldn't stop myself from saying, imaging the nightmare race of so many snakes scattering across the cone shaped hill. "Are they still there?"

"No," said Jim, "and we haven't seen one up there for years, though I did almost step on a rattler last year when I was taking a short cut across one of the vineyard blocks on the other side of 101." He gestured in the direction of the freeway, and then laughed. "I don't take short cuts though the vines anymore. Now I just stick to the dirt roads."

"Another interesting fact about Rattlesnake Hill," interjected Ned, "is that originally it was only planted with vines half way up. Dad found the original mounds when he was putting in the irrigation system for the new vines."

"Yes," nodded Jim. "The story was that the reason they only planted half way up the hill in the old days was because the mule died and they couldn't get the water to the top."

"Interesting," I said, thinking how difficult it must have been to farm in the old days when new vines needed to be watered by hand rather than the modern drip irrigation of today. "Do you have to use much water for these Zinfandel vines?" I pointed at the rows of vines in front of us.

"The old vines don't need much water at all," said Jim. "They were all dry-farmed, meaning that after giving them a few buckets of water when they were first planted, they were allowed to send their roots deep into the soil and find their own water. This soil has lots of clay so it retains water quite well." He paused before continuing. "However when we put a new vine in, we need to irrigate it so we've hooked the whole vineyard up to drip irrigation."

"Would you like to see our Petite Sirah block?" asked Ned.

As we walked towards the block, they told me more about their background. Jim went to CSU-Chico to study Animal Science and then took a job flying cattle, sheep, horses and other livestock to many different countries around the world. Eventually he became tired of traveling and returned home to Sonoma County where he married Julie Seghesio, had Ned, and then settled into farming the family vineyards.

Ned's background was similar in that he studied Fruit Science at Cal Poly in San Luis Obispo, and then worked for several years selling vineyard equipment before joining the family business. Today they hold the titles of Vineyard Manager and Viticulture/Grower Relations Manager, respectively, and oversee more than 350 acres including budgeting, managing employees, and grower relations.

As they spoke it was easy to see the strong connection between father and son. They both appeared happy, content and relaxed talking about their vineyard work. The manner in which they both listened to one another, and the ease with which they supported and added to each other's comments, showed the strong vein of respect that ran between them. I thought how rare it was nowadays in America to see a father and son working so closely together and still farming the land that was passed down from generations since the late 1800's.

"Here's the Petite Sirah block," said Ned.

Rattle Snake Hill at Seghesio Home Ranch

We had walked past all of the buildings and were on a narrow dirt road that crossed over a creek lined by small trees and bushes. The sunlight glinted off the tiny stream of water that trickled over rocks and tufts of long green grass. I realized that in other years that were less dry the creek would probably run much faster and wider.

Ned pointed beyond the creek to a group of vines on the left side of the road. They were huge shaggy creatures with long thick cordons stretched out on a VSP trellis system. Tiny green leaves with rose-colored tips sprouted along the dark bark, springing forth from the pea-sized buds that had recently awoke from their winter sleep.

"Interesting spacing," I said, noticing the vine trunks were more widely spaced between one another than across the rows.

"Yes," agreed Jim. "This is 14 acres of Petite Sirah on 6 x 10 spacing. Our spacing varies by block. "Pet" is prone to rot, so it is better to stretch the fruit out along the cordon as far as you can and give it lots of light. Many people don't like planting Petite Sirah because it can be so difficult, but it certainly is a great blending grape with Zinfandel."

I nodded, recognizing the shortened name of "pet" for Petite Sirah that many winemakers and viticulturists in California used. I also remembered the taste of many delicious Zinfandel and Petite Sirah blends in which the latter grape added tannins and color to the more lighter bodied Zinfandel. The trick was in not adding too much "pet," because it could overpower a wine due to its massive tannins and inky black color. "Tannin management" was the issue most winemakers associated with Petite Sirah, and apparently rot issues were its bane in the vineyard.

"Beautiful cover crop," I commented, nodding at the knee-high carpet of thick green grasses and tall beans topped by tiny pink and yellow flowers that spread lushly down the middle of the rows.

"It's a mixture of legumes, peas, bell beans and vetch," said Ned. "We plant the seeds in all the vineyards right after harvest, and the winter rains help it grown. Then in April, we spade it under to fertilize vines."

As I stood there in the vineyard I could detect a faint sweet fragrance wafting from the cover crop, and as I looked across the hundreds of vines with their newborn leaves and the thousands of flowers between the rows, I felt I was immersed in spring. The sky blazed blue overhead and I could see a hawk circling lazily in the distance. Reaching out to touch one of the long cordons of a Petite Sirah vine, I saw a small orange ladybug crawling along the dark bark. A good omen, I knew, because a ladybug was a sign of a healthy vineyard.

"I see you have beneficial insects," I said, nodding towards the ladybug.

Ned glanced over from where he was suckering a vine by gently rubbing off unwanted leaves that were growing on the trunk. I was also itching to do the same, a habit that must have become ingrained from tending my own vineyard, but I didn't want to accidently sucker the wrong leaf in this historic vineyard.

Ned smiled and nodded. "Yes, the cover crop is a great home to beneficial insects such as ladybugs that will eat mites and other "bad bugs" that may try to harm the vines."

How lovely, I thought, that this beautiful cover crop played so many important roles. Not only did it prevent erosion and serve as fertilizer, but it also protected small creatures and added to the beauty of springtime in the vineyard.

"Do you see the *Dr. Pierce's Medical Supply* advertisement on that old barn over there?" asked Jim.

I looked across the block towards where he was pointing, and could just make out the faded white lettering on the side of a weathered grey barn. It was a famous old-time ad that was visible to travelers driving along 101 towards Healdsburg.

"That's the edge of this property to the South," he continued, "but it keeps going up that hillside there and over the top to the West. We have blocks of different types of grapes there, including some Barbera and Aglianico. We also have some Carignan mixed in with the Zinfandel as a field blend. Come on," he continued, "let's walk over and look at our oldest block of Zinfandel on the other side of the house."

As we walked back towards the house, I could feel the sun beating down more warmly and wished I had remembered to bring a hat. Seeing me shield my head, Jim kindly suggested we look at the vines from the porch of the old house.

"You can actually see many of the old vine Zinfandel from this corner of the porch," Ned said as we perched ourselves on the edge of the ornate white wood railing. "They said my great-grandmother Angela used to set up a table in this corner and churn butter."

"It's a great place to sit," I said, enjoying the coolness and the shade of the porch roof over my head.

Ned and Jim Neumiller, Vineyard Managers

"These are the oldest vines in this area," Ned continued. "They were planted between 1895 and 1910, so some are over 100 years old."

We gazed out from the porch at the ancient gnarled beauties surrounding the old house, and I marveled at all those vines must have witnessed over the years. I could imagine the many conversations on this porch, and the comings and goings of countless visitors over the years.

The vines were pruned in the old European fashion of a *goblet* close to the ground. Standing about 3 feet tall, they all sported a wide base and thick arms spreading in multiple directions like a circular candelabra. Small green leaves sprouted from the ends of each thick arm.

"So why are you pruning these vines so close to the ground?" I asked.

"We've gone back to the old systems of trying to sculpt the vine to be a wheel or round," said Jim, "so the clusters are spread around the trunk, rather than climbing vertically like a ladder, as some people prune old vine zin. With the ladder, the fruit gets too tangled up and often overlaps. I think the old timers had it right and that head-pruned zin should be shaped like a wheel."

Ned nodded in agreement and picked up the conversation. "Pruning head-pruned vines is more challenging that VSP cordon. With cordon you generally just prune to one or two shoots per spur, but with head-pruned you have to analyze it. You have to picture what the vine will look like when it is green and filled with grape clusters." He held his hands in the air and drew a circle to demonstrate. "You have to make sure it won't get tangled up. It takes more decision making and experience to prune a head trained vine."

"That's for sure," said Jim, "and we want to make sure we don't have rat nests."

"I beg your pardon," I said in surprise. "Rat nests!"

Jim grinned. "That's what we call it when the grape clusters get all tangled up with each other as Zinfandel is apt to do if not pruned correctly."

"So how many buds do you leave per spur?"

"On these old head pruned vines we usually leave seven to nine spurs with two buds each," responded Jim. "We get lower yields, but the fruit is exceptional."

"We also prune carefully with our newer Zinfandel that is on VSP cordon," chimed in Ned. "I know it may sound weird, but we have started pruning to a single bud instead of two. This keeps the clusters from getting tangled up on the cordon as well. Even better, we find we have to drop less fruit now, and the vine is more balanced and gets ripe more evenly."

"Very clever," I said, impressed with the innovation taking place in the vineyard.

"We also have stopped following the rule that says all spurs must be pointed up," continued Ned. "We find it we let some grow down or sideways, that also helps alleviate the rat nest problem. So on a cordon we still have five spurs per foot, but some may be pointing down and others pointing up, so they are less likely to get tangled."

"But what do you do to protect these old vines from sunburn," I asked. "I'm sure it must get very hot in this part of Alexander Valley in the summer, especially given how hot it is today in March."

"You are right. It gets quite hot here, so we leave some extra shoots on the northeast side of our vines where they get a lot of afternoon sun," explained Jim. "We call this sacrificial fruit, since we can't use it because it is cooked, but it serves to protect the rest of the vines from sunburn."

"Sacrificial fruit," I exclaimed. "You certainly have a lot of colorful vineyard terms that I haven't heard of before." I glanced over at the vines, longing to get closer to them despite the hot sun. "Do you mind if we go over there now?"

Ned jumped up to lead the way, and we left the porch to walk over to one of the closest vines. It had ancient twisted arms that spiraled upwards making me think of a carved wooden octopus. Ned immediately began suckering the vines, deftly pulling off shoot that weren't needed and would take away energy from the tiny clusters that were not yet formed.

I reached out a hand to touch the bulging arm of a vine and found the bark warm under my fingers as I knew it would be. I marveled at all of the baby leaves and tendrils that I knew would grow into long shoots and drape over the vine like a lampshade with long hair. To some extent this would be helpful as it would shade the clusters from sunburn, but too much shading could prevent the grapes from ripening evenly.

"So how do you hedge these vines?"

"Now that's an interesting question," said Jim. "We've had a tizzy of a time trying to figure out how to trim the shoots, because you can't do it with a mechanical hedger because each vine is different, and pruning sheers are too slow."

"We tried machetes at first," said Ned, "but they were too aggressive."

"Then we tried car antennas, but they still were still to hard on the vines," continued Jim. "Finally I decided to go cut some willow branches down near the Russian River. Well, at first they were too wimpy, but then I dried them on the driveway overnight, and the next day they worked fine. So that's what we use to tip all the old vine zin." He motioned with his hand, pretending to wield an imaginary willow switch over the vines.

"You hedge with willow branches," I said amazed.

"Sure, why not?" Jim grinned. "It works!"

Vineyard Specifics for Home Ranch

As we continuing discussing the vineyard, Jim and Ned provided more technical information. The *190 acres* of Home Ranch includes 6 major *varietals*: 63 acres of Zinfandel with some field blend Carignan, 14 acres of Petite Sirah, and 1.5 acres of the old Chianti field blend. The remainder includes blocks of Sangiovese, Barbera, and Aglianico.

The *soil* is primarily clay loam with volcanic rock and red shale. According to Jim, "It's amazing when you dig down. In some places we find gravel, sand, big rocks, but other parts are loamy clay. Most of our old Zinfandel is on clay soil, which retains water pretty well. Therefore, it was easy to dry farm them in the old days." The *elevation* of the vineyard ranges from 205 to 500 feet above sea level, with Rattlesnake Hill topping around 500 feet high.

Temperature ranges from the mid 90's in July and can drop to the mid 50's at night. "It's usually always sunny here," said Jim, "and we have very little to no fog. However we do have a slight breeze that springs up in afternoon coming from around Rattlesnake Hill and that really helps." *Rainfall* averages 38 inches.

Vineyard Specifics for Home Ranch

Total Vineyard Acres	190 acres
Varietals	Zinfandel, Petite Sirah, Carignan, Sangiovese, Barbera, Aglianico, old Chianti field blend
Soil	Clay loam, volcanic rock, red shale, gravelly loam
Elevation	205 to 500 feet above sea level
Average Temperature	Summer temperatures average mid 90's F for a high and mid 50's F as a low. Rainfall is usually around 38 inches per year.
Rootstocks	St. George, 101-14, 03916, Boemer
Clones	Eduardo's Clone and Chianti Station Clone
Sun Exposure	East to west row orientation
Spacing	Varies from 6 x 3, 6 x 4, 6 x 10, 8 x 7, 7.5 x 7.5
Trellis Systems	Bush vine and VSP

Rootstock is St. George on the older vines, and they also have some 101-14 and 03916 in other parts of the vineyard. "For the new Zinfandel vines," stated Ned, "we are using a rootstock from Germany called Boemer. It is useful in preventing nematods, and we like it better than 03916, which is too vigorous and doesn't work well with Zinfandel."

Clones include the unique Eduardo's clone for the old vine Zinfandel, and it is also being used for budwood on all newer plantings of Zin. In the Chianti Station Block planted in 1895, they have also discovered that they have a unique clone.

"That block is a field blend with the Italian white grapes of Trebbiano and Malvasia mixed in," reported Jim. "The Sangiovese vines produce a very small berry which makes incredible wine, but we found that if you don't mix in the white grapes it is not as good."

"Interesting," I said. "So you are using the original Tuscan field blend recipe for making Chianti. What is the Sangiovese clone?"

"That's what is so fascinating," continued Jim. "We're not sure, but we know it is a very old clone. Several years ago we traveled to Tuscany and took photos of the leaves and clusters to the Sangiovese research center. The Italians looked it up and said it was a very ancient clone, and asked if we could provide cuttings. However we had to say no because it hasn't been cleared for transport out of the US."

"But there are many clones of Sangiovese," I said."

"Yes, apparently somewhere between 400 and 600, because each Tuscan village seemed to have its own clone. The Italians are doing a lot of DNA testing on Sangiovese to try to find the best clones to use."

"So what do you call yours?"

"Right now we call it the Chianti Station clone," smiled Jim.

The *orientation* for most of Home Ranch vineyard is east to west, and the *spacing* varies widely. Much of the old vine Zinfandel is on 8 x 7 spacing, and the Chianti Station block is 7.5 by 7.5, whereas the Petite Sirah is 6 x 10. "We used to have seven different spacing configurations in this vineyard," said Jim, "but we've managed to narrow it down to about five."

"However," added Ned, "we been experimenting lately with variable spacing of 6 x 4 and 6 x 3 based on the soil type. If the soil is more gravely and sandy we create a narrower window." He put his hand in the air and sculpted out an imaginary window.

"A window?" I said, puzzled.

"Yes." Ned smiled. "If you imagine the space between vines as the width of a window, it helps. A vine naturally wants to fill up the window by growing into it. If the window is too wide, the vine will still try to fill it, but won't be able to produce quality grapes. They especially can't fill it in poor quality soil. Therefore, you want to make the window, or spacing, smaller so the vine stays in balance. Make sense?"

He paused to look at me and I nodded.

"So," Ned continued, "we test the soil to see how wide the window should be. If we find the soil is too poor, we will actually make a smaller window between vines – even in the same row. We now have some vines that are 6 x 6 spacing and then 6 x 3 in the same row! It's working pretty well, and since we hand prune and harvest everything, it doesn't create any extra work."

"Wow," I said. "Now that is custom grape farming!"

Farming Methods at Home Ranch - Sustainable

Home Ranch Vineyard is *Fish Friendly Certified* and they are using sustainable farming methods but have not yet pursued other certifications. Natural *fertilization* is derived from spading in the cover crop each spring. In addition a combination of organic and non-organic fertilizers are used as needed with application through drip hoses. For *weed control*, they mow and spray Round-up under the vines.

Canopy management includes suckering in the spring, and hedging the tips of the vines in early summer using the willow switches. They drop fruit after verasion in July and August, especially focusing on discarding any rat nests. "We do multiple passes with the Zinfandel, usually about three times," said Jim, "to make sure we don't have any uneven ripening issue."

Harvest is usually in September and October, and a new cover crop is seeded immediately after harvest. Pruning takes place in November and December, but in some places on the property they will wait until March to avoid frost danger. After pruning the dead shoots are removed from the vineyard and composted.

Disease control primarily includes watching for powdery mildew, but it is not usually a large problem because of the sunny and drier climate. *Pest control* is also not an issue, except mites for which they have to spray. Nematods are controlled with the Boemer rootstock. Wild turkeys are an occasional hindrance in that they eat some of the grapes. They have owl boxes and raptor perches in the vineyard to help control rodents.

"What about deer?" I asked.

"Deer are not a problem," reports Jim with a grin on his face. "We had a consultant tell us to find their trails and then build a fence 25 feet on both sides. We did this, and it worked like magic. The deer came up to the fence and thought they couldn't come through, so they left." He paused, and then continued. "Wild hogs on the other hand are smart. They will go to the end of the fence, then come into the vineyard and eat the grapes."

Farming Practices at Home Ranch

Certifications	Fish Friendly Certified, Sustainable farming practices
Fertilization	Cover crop, organic & non-organic fertilizers as needed
Weed Control	Mowing, Round-up
Canopy Management	100% suckered and pruned by hand
Disease Control	Organic and traditional fungicides to prevent powdery mildew
Pest Control	No major issues, using owl boxes; Boemer rootstock to control nematodes
Irrigation	Drip irrigation
Technology	Weather stations, neutron probes, petiole analysis, digital imagery, and crop yield software
Harvest Measurements	Winemaker determines by taste and lab measurements

"You have wild hogs around here?"

"Yes," Jim said with emphasis. "And they are ugly things and make a mess. They get into a block and it looks like someone has gone in and torn up the soil."

Since the vineyard was originally dry-farmed, not much *irrigation* is necessary except during very hot weather in June and July, and if they are installing new vines. "We like to give the new vines short drinks," said Jim. "Not very much, because we want them to push their roots deep. We also take a lot of the fruit off new vines for the first ten years to encourage deep roots."

Technology in the vineyard includes weather stations for frost alerts, as well as neutron probes to test for water needs. Petiole analysis is conducted several times a year to determine if nutrients many be needed. "We also use a special software called Icrop Track," reported Ned, "that gives a yield estimate, as well as hire an independent Pest Control Advisor group not associated with any chemical company, to scout the vineyard to see what is needed. We include their input in the software system."

"And we use digital imaging," added Jim, "to analyze color and variations in the different blocks. But I find that if I climb to the top of the hill and look down at the vineyard, I can see by the color of the leaves if there are any issues, and if we need to irrigate. If I see some yellow-brown in a section of the vineyard, I know we need to do something."

Harvest measurements are determined by sugar samples that are collected and taken to the winemakers for analysis. "Instead of picking a plastic bag full of individual grapes as in most vineyards," explained Jim, "we have to pick a five gallon bucket of clusters due to potential uneven ripening with Zinfandel."

Jim and Ned monitor the vineyard very carefully as it gets close to harvest time. "I've been doing this for so many years," explained Jim, "that I look for certain things. For example, when I see the quail jumping on the wire to eat the grapes, I know it is about time to pick. I also look at color of seeds, and how easily they release from the pulp.

With Zinfandel, if the pulp releases right away, then it is ready." He stopped for a minute, grinned, and then kicked the soil with his boot. "However, with the Sangiovese, I just kick the vine. If some of the berries fall off, then it is ready."

Economics of the Vineyard

The average grape *yield* for Home Ranch Vineyard is 3 tons per acre, however there are a few blocks that may achieve 4 tons per acre. Therefore with 190 acres, the *total average tons* equals 570 per year.

Farming costs per acre average $7,000 per year. When they do put in a new block of vines, cost of development averages $24,000 per acre. According to Jim, "We are third, fourth, and fifth generation of farmers on this land. If we decide to pull out a block, we let the land rest for three or four years before installing a new vineyard."

Economic Viability of Vineyard

Average Yield	3 tons per acre
Total Average Tons Per Year	570
Costs Per Acre	$7,000 per acre
Revenues	Not specified
Economic Health	Very good

Part of the farming costs goes to employ 22 full-time vineyard workers, as well as an additional 35 part-time workers from Mexico. "Some of our full-time vineyard staff has been with us for 15 to 30 years," said Jim proudly. "For our part-time staff, we participate in the H2A program that allows us to bring the same 35 workers each year from Mexico. They arrive in April and depart again in November. We provide housing for them and a pay rate of $11 per hour. We are very happy with this workforce, because we trained them eight years ago and now they return each year. Before that we had over 175 part-time workers and many would quit after a few days on the job. It was frustrating and expensive to have to rehire and train constantly. Therefore we are much happier with this solution, and the workers are all very skilled."

In terms of *revenues*, it is difficult to quantify exactly, but the best fruit from Home Ranch vineyard goes into Seghesio's vineyard specific wines, such as the Home Ranch Zinfandel that retails for $50 to $60 per bottle and has achieved high scores from wine critics. The other varietals, such as the Home Ranch Sangiovese, Barbera, and Aglianico, have bottle prices ranging from $30 to $38. Fruit that doesn't meet the high standards of these special bottlings is usually blended into the Seghesio Sonoma County Zinfandel that retails for $24 a bottle, and has a production size of around 115,000 cases.

In terms of economic healthy, Home Ranch Vineyard appears to be performing admirably. According to Jim, "It is very profitable."

Soul of the Vineyard - Contentment

When I asked Jim and Ned to describe the vineyard in one word, we were standing on the porch of Eduardo and Angela's Victorian house with the vineyard spread around us. Ned leaned back against the wall of his great grandparents house and looked thoughtful. Finally, after a few minutes, he leaned forward with an earnest expression his face.

"Contentment, is the word I would have to use," he said. "Each day I get to kick the same dirt clods as my great grandfather did. I walk the soil of this vineyard and I feel honored to have the opportunity to work here." He paused and looked out at the old Zinfandel vines for a minute before continuing. "Yes, I would have to say that I tend the vines with contentment."

As Ned spoke, Jim nodded his head several times in agreement. "Yes, I love it here," he said with passion in his voice. "I love being in the vineyard with all the variety of work to do each season. I know it so well. It has everything I need to feel alive. As you know I used to work in the winery for a bit. I did everything there from cleaning tanks to bottling, but I prefer to feel the dirt under my feet."

Ned and Jim Suckering Vines

When we moved onto the topic of what they enjoyed most and least of working in the vineyard, their views were divergent. "A pro for me is working outside," said Jim, "and a con is that you can't control Mother Nature. However, because I know you can't control her, I don't worry about it. I just take each season as it comes."

"I'm the opposite," chimed in Ned. "I know it may sound strange, but I get a kick out of the challenge of gambling with Mother Nature. It's up to me to observe and make a decision on when to harvest, and how to beat Her. In the end, humans decide when the grape is ripe."

As I listened to their viewpoints, I was surprised at how different they were. Up until then they spoke in one accord and seemed to almost complete one another's sentences. Now Jim, the father, with many years of experience in the vineyard, admitted that he had accepted the whims of Mother Nature, whereas the younger Ned professed to enjoy the challenge of besting Her. Could they be opposite perspectives of the same coin, or were their responses ones that depended upon age and experience?

I thought of the thousands of vineyards around the world, and how for centuries the vineyard manager patiently tended the vines only to have frost, hail, or drought destroy the crop in certain years, whereas other years provided a bountiful harvest. In the end, how is it possible to control the whims of Nature? On the other hand, new technology in the vineyard has brought about many positive changes such as heaters and sprinklers to combat frost, overhead metal nets to temper hail damage, and new drought resistant rootstocks to assist with water shortages. Has Mother Nature finally met her match with Science?

"So if you enjoy the challenge of wrestling with Nature," I asked Ned, "then what do you least enjoy about working in the vineyard?"

Ned grinned. "Pretty much everything that takes me out of the vineyard, such as meetings, negotiations with growers, and budgeting."

"And what has the vineyard taught you over the years?"

Jim jumped in to answer the question first. "Things have changed so much over the years. When I first came to work here we were in the bulk wine business and we would pick the grapes at 22.5 brix. No one worried too much about quality; it was more about the quantity you picked and sold. Now it is all about quality and picking at the ultimate ripeness. For example we are focused on preserving the old vine Zinfandel because there is a market interest in them, even though they produce less fruit when they reach that age." He stopped and gestured at the vineyard blocks across Highway 101. "And the Chianti Station block, for example, we are preserving that to save history."

"I agree," said Ned, "we are very much focused on quality now, but I wasn't working in the vineyard when we were making bulk wine so I don't remember that. What I do remember is what my grandfather says: 'Prune me poor and I'll make you rich.'"

"Prune me poor and I'll make you rich?" I repeated. "What does that mean?"

"It's all about balance," Ned said. "If you prune off 96% of the vine – prune me poor, you will be rewarded with high quality fruit, which will make you rich. In the end it is about balance and patience."

As I drove away, I could see Jim waving at me in my rear view mirror and Ned continuing to gently sucker an old Zinfandel vine. Their words echoed in my head – patience, balance, contentment and happiness to be working the land. What a wonderful legacy for one of the original Italian immigrant families to share with others.

Glancing back once more as I drove down the frontage road, I could see the vineyard growing smaller. The old Seghesio homestead slumbered in the sun, and it seemed as if the vines wrapped themselves around it like a warm blanket of contentment.

Signature Wine - Seghesio Home Ranch Zinfandel

A week later I drove to Seghesio Winery in Healdsburg to meet with winemaker Ted Seghesio. As I entered the tree shaded parking lot and looked at the long building with a tasting room at one end, offices in the center, and the winery equipment the back, I remembered that this was the old Scantana winery that Angela has purchased. Today it was the new Seghesio Family Winery managed by her grandson Ted, who had a reputation as a brilliant winemaker and visionary leader for transforming the family business from bulk to high-end wines.

I was ushered into a private tasting room upstairs with a long oak table set with two iron candelabra. As I was invited to take a seat, I noticed there was a bottle of 2011 Seghesio Home Ranch Zinfandel in front of my place setting, along with a set of crystal wine glasses and spittoons in the shape of small clay amphorae. It was a few minutes before Ted arrived so I had time to look at the many awards on the walls, as well as a colorful slide show of the various Seghesio vineyard properties.

"Sorry I'm late."

Glancing up I saw a very tall, lean man with a full head of white hair and piercing light blue eyes. He was dressed in a navy sweater with blue jeans, and could have been a model for a Ralph Lauren advertisement.

"Nice to meet you."

We shook hands, and Ted took a seat on the opposite side of the table. After some casual conversation about the weather, I pointed at the slide show on the wall behind Ted. "I recognize that photo of Home Ranch Vineyard. Did you grow up there?"

"No," he said. "Pete's side of the family lived there. My grandfather was Authur, and my father was Ed. However we did spend a lot of time there when I was growing up." His blue eyes softened for a moment in memory. "My dad built a small camp ground on the bank of the Russian River near the Chianti Station block. We used to stay overnight and go swimming in the river. It was lots of fun."

"So where did you grow up then?"

"Here."

"Here? At the winery?"

"Yes," Ted smiled. "We lived in a house on the property. This is home." He spread his arms wide to encompass the room.

As the conversation continued, Ted described how the winery had changed over the years when they transformed from a large bulk operation with hundreds of cement tanks for fermentation and large redwood barrels for aging, to the smaller more upscale winery it was today.

"So why did you decide to switch from bulk to high-end wine production?"

Ted's expression softened. "My mother had a dream to have her own label, but we had been in the bulk business for so long, it was hard to change. We made wine and sold it to the giants of the industry, such as Gallo and Paul Masson. Then there was a surplus of grapes, and for three years in a row we couldn't sell our wine." He grimaced.

"So what happened?"

"That's when we decided to start our own label. We had lots of Zinfandel, and Paul Masson told us to make white zin. We started to do this in 1983 and were very successful, because everyone wanted to buy it. We sold that for several years until we hit the two great Zinfandel vintages of 1990 and 1991." He paused and his eyes lit up as he continued. "Those were phenomenal vintages, and suddenly everyone could see what red Zinfandel could do, so we started making serious wines. We were fortunate because my family had invested in some amazing old vine Zinfandel vineyards, such as Home Ranch."

"What, in your opinion, makes Home Ranch so special?"

Ted settled back in his chair and relaxed. It was obvious he was getting ready to discuss a topic he enjoyed. His voice was full of enthusiasm. "Four things. The soil is first. It is clay with volcanic rocks, which is a very unfertile soil. It has very little nutrients but the clay holds water, so it is great for dry farmed vines. So the soil creates devigored vines, which results in a small crop with fewer clusters each year." He paused to look at me to see if I was following. I nodded quickly.

"Second is the climate. Home Ranch is in a very warm location and Zinfandel likes warmth to assist with more even ripening. Even with the climate though, zin is notoriously difficult to ripen evenly. I think Zinfandel is the most difficult grape to work with because of this."

His comment made me remember clusters of Zinfandel I had seen hanging in Sonoma and Mendocino county vineyards over the past decade. They often looked like a long bag of multi-colored marbles with a mix of green, pink, red, and shriveled purple grapes all on one cluster.

It could be a maddening grape varietal, because while some grapes were over-ripe, others would still be green and vegetative tasting. That's why some winemakers wait until a large number of the berries on a Zinfandel cluster are shriveled, in order to insure the whole bunch is ready to harvest. The downside of this is the creation of high alcohol wines with ripe jammy flavors. Though many consumers like this style, there are others who find it to be out of balance.

"The third reason," Ted continued, "is aspect. Almost every afternoon at Home Ranch Vineyard, there is a cooling breeze that comes around Rattlesnake Hill and tempers the vine. Finally, I believe that our unique clone and the St. George rootstock play a key part in the high quality fruit coming off that vineyard. For example in 2010 when we had a heat spike that sent temperatures to 116 F, we lost some fruit because it burned, but the old vine Zinfandel and Chianti Block were fine."

His comment about the old vines reminded me of a common debate in the wine industry regarding whether or not old vines actually produced higher quality fruit than younger vines. "So do you think there is a difference in taste between your old and new Zinfandel vines in Home Ranch?" I asked.

"Great question," said Ted, leaning forward across the table. "You know we've planted new blocks of Zinfandel from the same budwood as our old vine zin, and we get great results, but it tastes different."

"How is that?"

"The old vine Zinfandel delivers an inherent berry and spice with good concentration, whereas the newer vines have a lighter mixed berry note and are not as concentrated. I think the new vines have different flavors because they don't have the deep root structure and history of old vines. They haven't been consistently devigoured over such a long period of time."

"Interesting," I said, thinking it was a very logical explanation for why older vines often delivered more concentration and complex flavors in wine.

"Why don't we taste it, and let me know what you think?" Ted poured wine into the glasses, and lifted his glass in a silent toast.

After toasting, I held my wine up to the dim light in the room, and could see that it glowed with dark ruby lights and purple depths. Swirling the glass for a minute, I then took a deep full breath so I could smell the aroma completely. Mixed berry compote filled my nose with top notes of raspberry, allspice and a hint of floral. Sipping the wine the first thing I noticed was a crisp acidity on the palate, most likely due to the cooler vintage of 2011. As the wine traveled around my mouth, I felt a pleasing velvety texture on my tongue and a hint of the graham cracker crust that the Home Ranch terroir was noted to give. Then blueberry and a brambly raspberry, and a slightly complex earthy note arrived. The tannins were light but still grippy, and the finish was long with a tart edge. It was a restrained Zinfandel, the style I enjoy, with good balance, though still a bit young to drink.

"It has 13% Petite Sirah in the blend," said Ted, "which helps bump the tannins up a bit and adds color."

"Very nice. I enjoy the elegance and balance of this wine. It's not your typical jammy high alcohol zin."

"We don't produce jammy wines," Ted said abruptly putting his glass on the table. "Jammy means sweet, and sweetness masks the distinct varietal characteristics that makes wine special."

"So how do you make these two different styles of Zinfandel?"

"Come on, I'll show you." Ted shoved his chair back and rose quickly to his feet. He ushered me through a door at the end of the room, and I followed him quickly down a long hallway and through another door. Suddenly we were outside in the bright sunlight, and I could see a row of small stainless steel fermentation tanks lined up under a large tin roof.

"This is where we made the Home Ranch and other specialty lots of wine," said Ted, striding rapidly over to the tanks. "These are six-ton open top fermenters."

"Impressive," I said, looking around at the vast array of tanks and a large bladder press. "Is there where you sort the grapes as well?"

Ted nodded. "We have a sorting table set up here during harvest, but sorting starts in the vineyard where we make multiple passes. We make sure all shoulders are removed from the clusters and the fruit is as ripe as possible without being over ripe. When the fruit arrives here at the winery, there are no green berries on the clusters. The grapes are destemmed, and then the sorting table crew picks out all the pink berries as well as any shriveled ones. We do, however, keep 10 – 15% of the shriveled grapes because we find it contributes to the spice characters in the wine."

Ted went on to explain that after sorting, the grapes are lightly crushed, put in the six-ton fermenters and inoculated with Fermrouge yeast. All varietals and blocks are fermented separately, and punched down, both mechanically and by hand, once a day in beginning and up to four times per day when fermentation takes off. The temperature is kept to around 85 F degrees, and fermentation generally lasts ten days.

"So since the Home Ranch Zinfandel is a field blend, do you ferment some of the Carignan and Petite Sirah together?"

"No," responded Ted. "Even though it is field blend, we pick it all separately because we want to make sure everything is at correct ripeness."

"How about adding acid?"

"We've never had to add acid to Home Ranch fruit until 2012. The acidity in the wine you just tasted was all natural."

I nodded, impressed by the fact that the 14.8% alcohol of the wine was perfectly balanced with good acidity and well-integrated French oak. "So no extended maceration before oak aging?" I asked.

Ted shook his head. "No, we've never really done extended maceration here on our Zinfandel. It's not really necessary. When fermentation is finished we use a bladder press, but I only use the free run juice for the Home Ranch Zinfandel. It goes into 25% new French oak barrels for 12 to 14 months."

He led the way into a temperature controlled room that was stacked to the ceiling with barrels. Immediately the toasty scent of spicy oak and wine filled my nose. I took a deep breath, enjoying the heading aroma.

"When I was growing up this room was filled with giant redwood tanks," said Ted. "They were almost as tall as the ceiling. We used them to age the bulk wine."

I nodded having seen some of the old tanks in other cellars. They were a tribute to a by-gone era when California wine was aged in the wood of giant Redwood trees, taken from forests that were now protected. Looking around I saw the cellar was filled with smaller 225 liter barrels made of French or American oak.

"Why French oak instead of American?" I asked, knowing that many Zinfandel producers use American oak.

"I prefer French oak because it respects the fruit," said Ted. "It is less aggressive than American oak, not as flashy and provides roundness and structure with hints of cinnamon and clove. It contributes to a more refined style of Zinfandel. However, we do age some Zinfandel in American oak as a stylistic choice."

"So what are the other differences in winemaking between the Home Ranch Zinfandel and the Sonoma County zin?"

"The Sonoma County zin is from all over the county, and we buy some of the grapes. We put it in large rotary fermenters, and it is aged nine months in American oak rather than the 12 to 14 months in French oak that Home Ranch receives." Ted stopped next to a stack of French oak barrels and ran his hand over the smooth wood surface. "And of course we make only 30,000 cases of our vineyard designate wines, while Sonoma County Zinfandel totals around 115,000."

As we left the barrel room and made our way to the visitor's center Ted provided more details about the aging process for the Home Ranch Zinfandel. While in barrel it was topped monthly and racked twice. Malolactic fermentation occurred in barrel and was induced. It was blended and allowed to marry in tank for one month before bottling.

"So what is your opinion on how long Zinfandel can age?" I asked.

Ted smiled. "Zin ages beautifully," he said. "If fact I'd prefer that people let my Home Ranch Zinfandel age for 3 to 6 years in bottle before opening, but that rarely happens. However, this zin can easily age for 10 to 15 years and still taste beautiful."

He paused for a minute, as if lost in thought, and then continued. "You know that is one of the lessons of the vineyard. Time – time and patience. The vineyard has been in my family for over a hundred years, and it makes me feel good to know that my nephew, Ned, and his father, Jim, watch over it every day."

Chapter Six

Stag's Leap Vineyard (S.L.V.)

Stags Leap District AVA, Napa Valley

Neon-orange poppies appeared in clusters along the Silverado Trail as I drove to Stag's Leap Vineyard on a breezy April day. The sky was a clear porcelain blue, not yet as vibrant in hue as it would become in the hotter summer days ahead, and the temperature matched the spring season with a moderate 68°F. Vineyards marched along both sides of the two-lane road, their foliage a pale-green blur with small leaves only a few weeks beyond bud break.

I drove slowly, just below the speed limit, so I could enjoy the expansive view. April is still a relatively calm month in Napa in terms of tourists, and the Silverado Trail, running parallel to crowded Highway 29 on the opposite side of the narrow valley, usually has much less traffic and so is a preferred route of many locals. However, despite the lack of traffic, my leisurely speed ended up irritating other drivers who sped up to pass me.

The turnoff to Stag's Leap Wine Cellars appeared suddenly, obscured slightly by large trees on a gentle turn in the road. The sign announcing the famous vineyard, home of the Cabernet Sauvignon wine that won the Judgment of Paris tasting in 1976, is small and tasteful. After parking my car underneath some trees in the shady parking lot, I entered the tasting room where I was scheduled to meet Kirk Grace, vineyard manager.

Since I was a few minutes early, I perused the merchandise and was delighted to find a large pile of George Taber's book, *Judgment of Paris*, detailing how an unknown wine brand from California, the 1973 Stag's Leap Wine Cellars Cabernet Sauvignon, had beat out Grands Crus Bordeaux wines in a blind tasting in Paris, France, in 1976. The results of the tasting were publicized by *Time* magazine and put Napa Valley on the world map as a top-quality wine region. Though I had read the book several years ago, I picked up a copy and was soon so engrossed in the chapter on Stag's Leap Wine Cellars that I was caught off guard when a tasting-room representative brought me a glass of Sauvignon Blanc to sip while I was waiting. The wine was fragrant and delicious, and I was enjoying both the beverage and the book when Kirk arrived.

"Hope you weren't waiting long," he said as he reached out a hand in greeting. He was dressed in the standard California vineyard manager wardrobe of blue jeans and work boots, complemented by a green checked shirt. He seemed to bring a breath of the outdoors into the tasting room.

"No, not at all. I was enjoying reading this book again."

Kirk glanced at the book and smiled. "Yes, great reading." I noticed that he had light-blue eyes in a tanned face, and the silver wings in his short, sandy hair gave evidence to his many years of experience, including managing the vineyards of both Sterling and Sinskey Wineries before moving to Stag's Leap Wine Cellars seven years earlier. He had a degree in Crop Science from Cal Poly and an extensive background in environmental science. A native of Napa Valley, his parents had founded Grace Family Vineyards, so Kirk had grown up immersed in local agriculture issues.

"Where would you like to start today?" he asked.

"S.L.V.," I said, using the shortened name for Stag's Leap Vineyard.

"Do you also want to see FAY Vineyard?"

"Definitely!" FAY was the original vineyard that had inspired Warren Winiarksi, the founder of Stag's Leap Wine Cellars, to plant S.L.V.

Kirk smiled. "Great, let's go."

History of Stag's Leap Vineyard

It was in the late 1960s that Warren Winiarksi had the epiphany that led him to plant Stag's Leap Vineyard. A former professor, he had been working in Napa Valley as a consulting winemaker and grape grower for several years, and was on the lookout for a property that he and his wife could purchase in order to start their own winery. One day he was visiting with Nathan Fay, who owned a small Cabernet Sauvignon vineyard in the Stags Leap District just north of the town of Napa off the Silverado Trail. Nathan offered him a glass of wine made from his vineyard, and it changed Warren's life.

According to an interview in *The Winemaker's Dance*, Warren reported:

> *"I was unprepared for the experience when I first tasted Nathan Fay's wine...when the perfume of it spread through the small room where we stood together...I recognized immediately that this was the kind of wine I wanted to make (p. 126)."*

Warren was transfixed by the experience and immediately began to look for property near Nathan Fay's vineyard. Fortuitously, the fifty-acre Heid Ranch next door was for sale, and Warren was able to purchase it for $110,000. He then set to work clearing the land of plum trees, and in the spring of 1970, he planted Stag's Leap Vineyard with two-thirds Cabernet Sauvignon vines and one-third Merlot.

Warren tended the vineyard carefully over the next three years, and then in September of 1973, when the vines were of sufficient age to make quality wine and the growing season had been nearly perfect, he hired a crew of pickers to assist with the harvest. Over a five-day period, they picked thirty-two tons of grapes from the vineyard, and Warren hired his good friend, André Tchelistcheff, to provide advice in the winemaking process.

The grapes were fermented in individual lots and aged in French oak. The final blend was 90 percent Cabernet Sauvignon and 10 percent

Merlot, and the wine was released to the market in July of 1975. There were a total of eighteen hundred cases produced, and it was named the 1973 Stag's Leap Wine Cellars Cabernet Sauvignon.

It was this wine that Steven Spurrier, a British wine merchant living in Paris, selected as one of six California red wines to compete in a blind tasting against famous French wines. Surprisingly, the Stag's Leap wine came in first, winning out against the great Bordeaux wines of Château Mouton Rothschild, Montrose, Haut-Brion, and Léoville-Las Cases. Even more amazing was the fact that all of the judges were well-known French experts in wine and cuisine.

When the results were announced, they "had a revolutionary effect, like a vinous shot heard round the world," wrote Barbara Ensrud in the *Wall Street Journal*. Overnight, Napa Valley became known as a famous wine region, and Stag's Leap Vineyard was recognized as one of the most legendary plots of land in the world.

In 2007 Warren retired and sold the winery and vineyard for $185 million to Ste. Michelle Wine Estates in Washington in partnership with Marchesi Antinori. Today the winery continues to produce the S.L.V. Cabernet Sauvignon from thirty-five acres of Cabernet Sauvignon and 1.5 acres of Merlot, though the original 1970 vines have been replanted.

Touring Stag's Leap Vineyard

Kirk ushered me into a large four-door silver pickup truck parked behind the tasting room, and soon we were driving down a small gravel road and into the front section of Stag's Leap Vineyard.

"As you may know," said Kirk, keeping his eyes on the road, "S.L.V. is currently thirty-six point five acres and includes ten different blocks that have been planted at different times over the years. Right now we are driving through some of the newer blocks, but I'm taking you to the back portion of the vineyard where vines were planted in 1972."

Gazing out the window, I was impressed with how green and healthy all of the vines looked. They were organized in straight rows with the trunks about three feet tall and long cordons on each side trained to a VSP trellis system. Bright, new green leaves flared on shoots that ranged from two to six inches in length. Between every other row, a carpet of short green grass with tiny yellow wild flowers spread out in long strips, while the opposite rows showed bare, brownish-red soil. It looked like a luscious field of striped saltwater taffy.

Newer Block in Stag's Leap Vineyard

Kirk steered the large truck around a curve in the road, and I noticed that we were closer to the craggy knoll that rose tall and sandy colored above the back of the vineyard. I knew these were called the Palisades, and they were an impressive range of rocky ridges that gave proof to Napa Valley's volcanic past.

"We call this Block Four," said Kirk, stopping the truck. "This is currently the oldest section of S.L.V., and I had a chance to work with the founder, Warren, for a few years before he retired. Unfortunately, the 1970 vines that went into the wine that won the Judgment of Paris are gone due to old age and disease, but this section was planted in 1972, so it is still quite historic."

I jumped down from the truck and walked quickly toward the wide rows of big, shaggy vines. They stood with outstretched arms, and with the shorter green foliage of the season, looked almost like a group of eagles with wings unfurled, waiting to take off. A few of the vines had small holes in the wide trunks near the top of the cordons, which gave them the appearance of having a face.

"What caused these holes?" I asked.

Kirk smiled. "We actually have some squirrels that place acorns in these old vines."

"Does it hurt the vine?"

Old Vines in S.L.V. Block 4 with Squirrel Holes

"No, because the vine's vascular system is able to work around the damage, and the vine gets most of its nutrients from the roots. Besides, it is part of our program to support and protect wildlife."

Once again I was reminded of the wide diversity of wildlife that is attracted to vineyards and lives in a symbiotic relationship with the vines. Stretching out my hand, I touched the shaggy surface of the vine that had lived more than forty years and felt the warmth of the sunshine on the bark. The new leaves were a combination of pale- and medium-hued green, and tiny shoots hinted at the grape clusters that would form and bloom in a few weeks.

"I wish I had asked the guys to sucker before you came," chuckled Kirk. "These vines need to be cleaned up a bit, but we weren't going to do it until next week." Kirk was describing the tiny green shoots that were sprouting from different parts of the bark and would be taken off by the vineyard crew so the vine could focus all of its energy on growing the main shoots with the grape clusters.

"Makes me feel better that you didn't, because now I can see them in all their natural glory. Besides, now I don't have to feel as guilty that I haven't suckered my hobby vineyard yet."

Kirk laughed and then directed my attention to the ground. "So take a look at this soil," he said. "Do you see how it is slightly reddish-brown in color? This is because S.L.V. is primarily on volcanic soil that has poured down in alluvial fans from the Vaca range up there." He turned and pointed to the craggy Palisades behind us with the even steeper Vaca Mountain range beyond. "As you can see, this part of the vineyard is also on a slope, whereas the newer part of S.L.V. is on flatter land."

Glancing around, I realized he was right, and the vineyard was slightly higher where we were standing. The slope increased in height as the vines marched further toward stands of dark oak trees and the craggy mountains beyond. High above the vineyard were two large outcroppings of rocks with a narrow opening in the middle.

"So is that the famous cleft in the mountain where the deer story happened?" I asked, pointing to the rocky cliffs.

"Yes," Kirk grinned. "Apparently in the olden days, the Indians used to hunt deer up there, and their method was to chase the deer against the cliff face to corner them. However, one day, a very smart deer got to the top of the rocky cliff on the left, and instead of falling to his death, he leaped across the divide there and landed safely on the other side. So it became known as Stag's Leap."

"What a great tale!"

"Yes, so the settlers named this district Stags Leap, and later, when the winery was founded, the term was incorporated into Stag's Leap Wine Cellars."

Block 4 of S.L.V. with Cleft in Rocks Where Stag Leapt Across

"What about the other winery that is also called Stags' Leap," I asked. "Isn't that confusing to consumers?"

Kirk sighed. "Yes, it can be confusing, so we make sure to label our bottles with the names of the vineyards. Thus, you can find S.L.V. and FAY Vineyard bottles, as well as KARIA, ARTEMIS, and ARCADIA Vineyards. In addition, since 2009, they have added a seal to the labels stating that ours is the winery that won the 1976 Paris Tasting."

"Good idea." Glancing around at this very famous vineyard and thinking of the historic tasting in Paris, I realized that the vines must have been quite young to produce a 1973 Cabernet Sauvignon. "So if Warren planted the vines in 1970, and the first harvest was in 1973, then the vines were only three years old! Just babies!"

"Yes," Kirk nodded. "There are some people who say vines go through stages of growth where certain ages produce better quality wines. For example, current wisdom says that vines produce higher quality when they are very young and when they are older, but not as good during the troublesome 'teenage' years." He laughed. "Whether this is true or not is not clear, but it is a fact that the 1973 vintage was excellent in Napa Valley. So even though our vines were young, they produced a wine that won the Judgment of Paris."

His comments caused me to remember conversations with other winemakers in California, France, and Australia. They all seemed to have different opinions on vine age and quality. In the Rhône and Burgundy, some had the opinion that the vine should be at least fifteen or twenty years old for the best production, whereas a winemaker I know in Bordeaux believed that young vines ranging from three to five years often produced exceptional wines. At the same time, several California and Australian winemakers told me that older vines produced the highest quality but at lower quantity. There was no clear consensus, and perhaps this is because it may also have to do with the location, type of varietal, and clone. However, since the three-year-old S.L.V. vines produced an award-winning vine, there does appear to be some credence to the young vine theory.

"What is your viewpoint on vine age?" I asked.

Kirk looked thoughtful for a minute before answering. "I believe that a vine can produce the same amount of quality every year if you take care of it. Instead, most people try to up the tonnage as quickly as possible before the vine has established its individual architecture."

"What do you mean by establish architecture?"

"Well, each vine has to set its cordon, form spacing between spurs or canes, and develop its root system. This will determine what it looks like as it ages. The vine needs time to do this well without being forced to overproduce. For example, common knowledge states that a one-year-old vine should produce half a ton; a two-year-old vine should produce one and a half tons, and a three-year-old vine should produce three tons. Instead I think this should be allowed to happen more slowly."

I nodded at his explanation, recognizing the long-standing battle between producing both high quantity and high quality in a vineyard in order to achieve positive economics.

"Let's check out some of the newer blocks," said Kirk, and I followed him as he walked through the dusty soil covered with small rocks and crossed the gravel road to a block on the opposite side. Here the spacing was tighter at seven by four feet, and the vines were obviously younger with slimmer trunks and cordons. In a few spots, a slender new vine had been planted and was growing up the central pole in the VSP trellis system.

Kirk Grace, Vineyard Manager, Stag's Leap Wine Cellars

"I see you are planting some new vines."

"Yes. When the older ones die out, then we replace them," responded

Kirk. "We are trying to use the S.L.V. Clone where possible."

"The S.L.V. Clone? You have your own clone?" I knew that over time, grapevines mutate to create their own "personality," or character, that is distinctive and can be referred to as a new clone but still possessing the DNA of the mother plant. However, I didn't realize there was an S.L.V. Clone.

"We've tried to save the budwood from the older Cabernet Sauvignon vines and use it to graft new vines." He bent down to touch the young vine. "Since this vineyard has such an impressive history, we want to preserve it. We think the original budwood for the S.L.V. Clone was from Martha's Vineyard just south of Oakville and that it is related to clone two; however, the history of grape budwood selection during Napa's infancy is highly fractured, so take this with a grain of salt."

Gazing at the fragile vine climbing so valiantly toward the sky, I felt a lump of emotion as I realized it contained the same DNA as its famous grandparent vines planted in 1970. Those vines were destined to change the course of California wine history by creating a wine that beat out the great and famous châteaux of Bordeaux.

"Would you like to see FAY Vineyard? It is right next door."

Kirk's words woke me from my reverie, and we climbed back into the truck to drive a short distance over a dried creek bed and into a large swath of vineyard on flatter land on the other side. Kirk stopped the truck in the middle of the vineyard, and we jumped out to walk deep into the rows of FAY. I noticed that the dirt was darker brown in color and looked like rich cocoa powder.

"As you can see, the soil is different in FAY compared to S.L.V.," he said. "FAY is primarily clay soil and creates softer, fruitier wines. S.L.V. is volcanic, and we believe this causes it to grow wines with bigger tannin structure and darker, more complex fruit."

"Fascinating."

"We often use the analogy of water and fire. FAY is water because her soils are clay, and she produces softer wines. S.L.V. is fire because of its volcanic past and produces wines of bigger structure. This is also why we produce CASK 23, which is a blend of the FAY and S.L.V. Vineyards."

Looking back toward S.L.V., I could see that it was located to the south of FAY and stretched higher on the hillside with its distinctive red, rocky soil. However, FAY was much larger and stretched out over the flatter valley floor to the north of S.L.V.

"Another way to look at these two vineyards," continued Kirk, "is like a mini-Napa Valley. The rocky cliffs to the north are like Calistoga and hotter. We are surrounded by the Vaca range on the east and Pine Ridge Knob on the west, which is similar to the mountain ranges that surround Napa Valley. To the south, there is an opening—just like Napa Valley has—where cooling breezes from the San Pablo Bay blow in every afternoon around two or three p.m. This creates an ideal microclimate."

Vineyard Specifics for S.L.V. and FAY

As we toured the two famous vineyards set side by side, Kirk provided more detailed information. "FAY is almost twice the size of S.L.V., at sixty-six acres compared to thirty-six point five," he said.

In terms of *varietals*, FAY has 65.5 acres of Cabernet Sauvignon and a half-acre of Petit Verdot, whereas S.L.V. has 35 acres of Cabernet and 1.5 acres of Merlot.

The geological origin of the *soil* comes from the Great Valley Segment, which is a mixture of rhyolite and andesite composed of volcanic soil, gravel, rocks, and clay.

"There are many alluvial fans flowing into these vineyards," Kirk commented, "and one of the nice aspects is we have the perfect pH in the soil of both vineyards, which helps to create very balanced wines."

The *elevation* ranges from 85 feet in the flatter part of the vineyard and up to 220 feet on the higher slopes of S.L.V. Average summer *temperatures* range from 85°F to 94°F, with nighttime lows ranging from 50°F to 55°F. A unique aspect of the property is the breezes that sweep up from San Pablo Bay in the afternoons to cool the vineyard and help to preserve high acids and flavors.

Rootstock in the older blocks is St. George, with newer blocks planted primarily to 101-14 and 420 A. According to Kirk, 101-14 can be a troublesome rootstock "because it is not drought tolerant and we need to water more." There are also a few sections of the vineyard that are on 110-R and 3309 rootstock.

Vineyard Specifics for S.L.V. and FAY

Total Vineyard Acres	S.L.V.: 36.5 acres (35 Cabernet Sauvignon; 1.5 Merlot) and FAY: 66 acres
Varietals	Cabernet Sauvignon and Merlot (FAY Vineyard has Cabernet and a half-acre of Petit Verdot)
Soil	Great Valley Segment; composed of Bale Clay Loam and Boomer-Forward Felta Complex
Elevation	85 to 225 feet above sea level
Average Temperature	Summer averages from 85°F to 94°F, nighttime lows 50°F to 55°F. Breezes every afternoon
Rootstocks	Oldest sections of S.L.V. on St. George; newer sections on 101-14, 420A, 110-R, 3309
Clones	S.L.V. Clone (originally from Martha's Vineyard), Concannon clones 7 and 8, and also have clones 2 and 4
Sun Exposure	North/south orientation to receive afternoon sun
Spacing	Older sections on 12 x 8; newer sections mainly 7 x 4
Trellis Systems	VSP, plus experimenting with Guyot and Small *Y* on Moser Trellis

In addition to the S.L.V. Cabernet Sauvignon *Clone*, they are using the Concannon Clone, which is also known as clones 7 and 8 and has been certified virus-free. Other clones include 2, 7, and 4, the latter being

131

one that Kirk refers to as a "shy setter." He explains the reason they use so many different clones of Cabernet Sauvignon in very poetic terms. "The variety of clones creates diversity in the wine. It is like a bouquet of roses. You can have just one color of rose, and it is beautiful, but multiple colors provide more scents and hues."

Most of the vine rows are planted in a slightly northeast/southwest *orientation*, so they receive the afternoon sun, which is important because Stags Leap District is considered to be a cooler area than other parts of Napa Valley. *Spacing* in the older blocks is on the traditional 12 x 8 feet, whereas newer sections are primarily 7 x 4 feet.

Experimental small "Y" trellis in FAY Vineyard

The majority of the *trellis systems* in both S.L.V. and FAY are VSP with double cordon for spur pruning; however, they are experimenting in FAY with some unusual options.

"We are trying out cane pruning on low Guyot," explains Kirk, "to see what type of result we get. We are also experimenting with a Moser trellis with a small Y because it allows the vines to hang in a curtain over the clusters, which prevents sunburn and provides better air circulation." He points out an unusual-looking group of vines, which are trained to a high Y trellis system with the cordons twisted in a circle and then spread out in opposite directions to create a "curtain" when the shoots are longer. It reminds me of Riesling vines I saw in the Mosel region of Germany, which are formed into a heart shape to provide more fruit exposure and to open the canopy.

Farming Practices in S.L.V. and FAY Vineyards

Back in the truck, we continued down the road through FAY Vineyard and started making our way back toward the winery and tasting room. As we rounded a curve, I noticed on the right that a vineyard block had been uprooted, and the old vines were stacked in large piles that rose over ten feet high.

"Looks like you will be having a bonfire with those old vines?" I commented casually.

"Yes," Kirk responded, "we had to tear out those two blocks for replanting. We will burn the piles of vines, but is it is regulated by the Bay Area Air Quality Management District and must be done on permitted days."

I squinted at the large piles, trying to see the contents better. "Are some of the old vines still attached to trellis wire?"

Kirk glanced over at the piles quickly. "All treated wood and drip lines must be removed. The logistics of separating the vines from the wire is impractical, but once the piles are burned, the metal is recycled."

"So is this part of your *environmental program*?"

Kirk nodded, refocusing his eyes on the road as he maneuvered the truck up a small hill. "We are Napa Green Certified, as well as Fish-Friendly Certified. Do you remember the dry creek bed we crossed over when we left S.L.V. and moved into FAY?"

I nodded, remembering the curving creek bed filled with large boulders instead of water and surrounded by tall oak trees.

"Well, the creek is dry this year because we didn't have much rain, but often we have fish running in it." He smiled, and when he glanced over at me, I could see how pleased he was to be part of an effort to bring the native fish back. Certain types of salmon living in California streams have been on the endangered species list. Therefore, many vineyard owners and other farmers have adopted practices to allow the natural creeks and springs to flow again without using all of the water for irrigation. Remembering Kirk's background in environmental science and his fondness for wildlife, camping, and all things outdoors, I thought again how fitting this vineyard position was for him.

Farming Practices at S.L.V. and FAY

Certifications	Napa Green Certified, Fish-Friendly Farming
Fertilization	Natural cover crops, foliar sprays as needed
Weed Control	Mowing, Roundup as needed
Canopy Management	Sucker, deleaf, hedge, crop thinning (prune to two shoots per spur)
Disease Control	Wettable sulfur, careful pruning to prevent *Eutypa*
Pest Control	Minor issues with nematodes, leaf hoppers, mites, and coyotes
Irrigation	Deep-drip irrigation; rainfall average: 27 inches per year
Technology	Frost equipment, neutron probes, pressure bongs, petiole analysis
Harvest Measurements	Team decision with winemaker; usually 24° Brix

"Isn't Napa Green Certified similar to the California Sustainable Winegrowing Certification?"

"Yes," said Kirk, "We have many of the same sustainable practices in the vineyard, but the Napa County programs exceed some of the federal, state, and local best practices. S.L.V. and FAY Vineyards are both Napa Green certified."

He continued to describe some of their farming practices. For *fertilization*, they use a cover crop of clover and two grasses called "blando brome grass" and "zorro fescue," which are mowed during the summer and then disked into the soil each fall. When I asked why they only plant cover crops in every other row, Kirk explained that it is to let the soil rest, and then the rows are rotated every year.

"So this year we have cover crops in the rows you see," he waved his hand out of the window toward the vines, "and next year we will switch rows so that the ones with only soil will then have a cover crop. If we tilled the whole vineyard and there were no weeds, the vines would get too vigorous and out of balance."

"Do you ever have to add anything else?"

"We do a petiole analysis every year to see if there is anything lacking in the soil. If needed, we will apply foliar sprays. Sometime we also need to add gypsum, prosperous, and potassium. However, in general, the soil here is naturally well-balanced and doesn't need much extra fertilizer such as manure."

For *weed control*, they will mow and occasionally use Roundup when the weeds are especially strong. "We are allowed to use this as part of Napa Green Certified," Kirk explained. "I don't use it every year; just in years where the weeds are not desirable."

Canopy management involves suckering the vines in April, deleafing and removing laterals (extra shoots) if necessary in June, some crop thinning and shoulder removal of grape clusters in July before verasion, mechanical hedging in July, and harvest usually in late September. During the pruning season in January, they prune to two spurs per shoot along the long cordons, which stretch in both unilateral (one arm) and bilateral (two arms) directions.

"What amazes me about our vineyards," Kirk commented, "is how much personal human work we do. Each vine is still individually pruned and harvested."

Like most California vineyards, the only major *disease issue* is powdery mildew. In order to prevent this, they spray a wettable sulfur every ten to fourteen days. If the weather indicates a higher than average risk of powdery mildew due to too much fog or moisture, they will spray more often. *Eutypa* is an occasional disease risk, which is a fungal infection that may occur during pruning because of large cuts to the wood. At Stag's Leap they attempt to control this by only pruning when the weather is dry.

In terms of *pest control*, Kirk reported that the only pests were leafhoppers, mites, and nematodes.

"Wildlife eating the grapes is not a problem here," he said. "We have deer fencing so the deer can't get in. I did see a cougar come down the hill to the vineyard once, but he was probably just hunting rabbits."

Just then, as if on cue, we both saw a large jackrabbit under the vines in a block we were passing.

"Hey, there's a rabbit," I yelled.

Kirk laughed. "Yes, we have rabbits, but they are not really a problem. However, they do attract coyotes, and coyotes will eat grapes. We've seen them strip the vines."

"Wow, and that's not a problem?"

"No, because it rarely happens. In general, wild animals don't cause problems in the vineyard."

"What other types of wildlife have you seen here?"

"Bobcats and wild turkeys. In fact, I think there is a group of turkeys up on that hill over there." He slowed down the truck, and we both could see a group of turkeys foraging in the scrub oak halfway up a small hill beyond the vineyard fencing. They moved slowly with their long necks stretched forward and the sun shining on their dark-brown feathers.

"What about small birds and yellow jackets?" I asked, thinking of the battle I had in my small hobby vineyard with birds and yellow jackets eating the grapes.

"Birds are not really an issue here, though they are in Carneros where it is important to net the vines." Kirk smiled. "It is wonderful how many natural materials we can now use in the vineyard compared to when I first started working in vineyards twenty years ago. We used to have to get tested each year for the potential impact of chemicals we were using. Now we use much healthier and natural materials."

Drip *irrigation* is used at Stag's Leap when needed. As the average rainfall in Napa Valley averages twenty-seven inches per year, Kirk reported that they mainly irrigate in July and August but also in the spring when the buds are forming.

"I believe in watering more at one time, rather than just a little bit at a time," he says. "If the water goes deep into the roots, this causes them to grow deeper rather than curve up to the surface. We've also found that deeper watering helps the vines create more intense phenolics (flavors) at lower Brix levels."

Technology in the vineyard includes not only petiole analysis each year but also the use of weather stations, pressure bombs, neutron probes, and fans.

"These new fans are called 'Tropic Breezes,'" says Kirk, pointing to large white fans positioned strategically around the vineyard blocks. "They are quieter and more energy efficient and effective than our old gas fans and help prevent frost damage to the vines." He rubbed his forehead and grimaced. "I still remember 2008. It gave me nightmares. I was woken up twenty-two nights in a row with frost alerts and had to turn on the fans to save the vineyard."

In terms of *harvest measurements*, Kirk works in partnership with the winemaker and the rest of the crew to taste the grapes and conduct sugar samples. "I'm always asked for my opinion on the best time to harvest," he says, "but in the end, the winemaker makes the call. In general we harvest around twenty-four degrees Brix and usually bring in the crop around the end of September, but that will vary based on the year."

Vineyard Economics at Stag's Leap

The *average yield* for the old portion of S.L.V. is 2 tons per acre, but newer sections achieve up to 3 tons per acre. In general, FAY Vineyard achieves 3.5 tons per acre. Therefore, if S.L.V. is 36.5 acres with an overall average tonnage of 3, this equals 109.5 tons per year, and FAY with 66.5 acres and an average production of 3.5 tons per acre generally achieves around 232 tons.

Economic Viability of S.L.V. & FAY Vineyards

Average Yield	2.5–3.5 tons per acre
Total Average Tons Per Year	109 S.L.V.; 232 FAY
Costs Per Acre	$8,000–$11,000
Revenues	Not available
Economic Health	Positive

Annual farming costs per acre range from $8,000 to $11,000, depending on the amount of labor needed in the vineyard and based on the weather that year. According to Kirk, "We have outsourced our labor to Silverado Farms. All of our full-time workers got jobs there with full benefits, which makes them happy and me happier. Therefore, we have excellent people working in the vineyard who have been with us for years." An interesting statistic is the fact that they generally employ one full-time vineyard worker to care for fifteen acres. The average hourly wage is $12 to $15 plus benefits.

Stag's Leap Wine Cellars is also redesigning some vineyard blocks to save tractor time and fuel by using longer, straighter lines. "This way the tractor has to turn around less," explained Kirk. The average cost for new vineyard installation is around $25,000 per acre.

In terms of *revenues*, Stag's Leap Wine Cellars is part of Ste. Michelle Wine Estates, which is owned by Altria Inc., a publicly traded company. In 2013, wine revenues were $609 million according to the Altria annual report, but it is difficult to determine the exact percentage contributed by Stag's Leap Wine Cellars, as they are one of several wineries in the portfolio. Furthermore, their retail bottle price is much higher than some of the other wineries, with S.L.V. priced at $125 per bottle, FAY at $110, and CASK 23—a blend of the two vineyards— selling for $210. In addition to these estate-tier bottles, Stag's Leap Wine Cellars also produces a Napa Valley Collection for more moderate prices, bringing their total annual produce to around one hundred thousand cases.

"All of the fruit harvested in S.L.V. and FAY Vineyards is used for the estate wines," commented Kirk. "We do not sell any grapes." In terms of *economic health* of the vineyards, Kirk stated, "In general, things are pretty positive, but as a publicly traded company, there is always pressure to reduce costs and maintain quality."

Soul of the Vineyard—"Prettiest"

When asked to describe the vineyard in one word, Kirk replied, "That is easy. This is the *prettiest* vineyard I've ever seen." We were standing at the foot of S.L.V. Block 4, and he turned in a circle and pointed toward the top of S.L.V. with the tall craggy rocks and peaks rising valiantly behind. "I've worked in a lot of different vineyards in my life, and there is no place else as special as this. Look at the slope with the old vines, the oak trees behind, and those amazing mountains encircling it all."

When asked to describe the challenges of working in the vineyard, Kirk mentioned change management issues.

"Before coming here, I was working in an organic vineyard and wanted to bring some of those methods to S.L.V. and FAY," he explained. "However, it is not always easy to convince people of the benefits of that type of change. This is because they had been using conventional farming methods here for a long time, such as spraying to eliminate weeds and applying lots of water and fertilizer. I wanted to plant cover crops, do less tillage, and bring the soil back into balance. However, now, through collaboration and teamwork, we have created some positive changes."

When asked about the best aspects of working in the vineyard, a large smile lit up his face. "That's easy—the results." He turned around again and gestured toward the healthy green vines. "The soil is in balance, and the quality of the fruit is higher. The organic matter in the soil is higher, which is quite positive, and we have the ideal pH. I feel good to have helped make this happen."

As we come to the conclusion of our vineyard tour, I asked Kirk what working in S.L.V. and FAY had taught him. He walked over to a large, healthy vine whose cordons were spread lengthwise in both directions along the trellis. Small green leaves waved in the gentle afternoon breeze, and Kirk placed his hand on the bark of the cordon.

"Canopy management," he replied, running his hands gently along the cordon. "I came here with a lot of soil management knowledge, but over the past few years, I've learned a lot about canopy management."

"How did you learn? By taking classes?"

"No, the winery founder taught me."

"Warren Winiarksi?"

"Yes, he was still here the first year and a half when I started. He had so much knowledge of the vineyard and was very meticulous in taking care of it. He spent hours out here with me, explaining how to manage the canopy to produce the optimal fruit. We focused on reducing sprawl, checking leaf count, and regulating the laterals so the clusters had the right amount of nutrition and sunlight."

As he spoke, I had a vision of Warren standing next to Kirk in the vineyard, inspecting each vine meticulously and counting leaves. It was a beautiful image of mentor and mentee, and a great metaphor for how vineyard care and wisdom is passed down through the ages. Now Kirk can share the wisdom with his workers, who will hopefully pass it on to the next generation of vineyard caretakers.

When we were driving slowly out of the vineyard, with the wheels of the truck crunching gently over the gravel road, I glanced back and looked one more time at the famous vineyard stretched out in a green tapestry with the craggy beige mountains rising behind it. A hawk circled lazily overhead, soaring on the afternoon breezes that tickled the tops of the vines, and in the distance, it seemed as if the shadow of a stag did, indeed, leap across the prettiest vineyard in Napa Valley.

Signature Wine: S.L.V. Cabernet Sauvignon

Since winning the Judgment of Paris tasting with their first release of the 1973 S.L.V. Cabernet Sauvignon, Stag's Leap Wine Cellars has continued to produce this wine every year. Obviously there is vintage variation, but in general the vineyard continues to produce high-scoring wines. This was also the case with the 2009 S.L.V. Cabernet Sauvignon, which received ninety-one points from *The Wine Advocate*. The 2009 CASK 23, which is a blend of S.L.V. and FAY, actually scored higher, receiving a ninety-four from both *The Wine Advocate* and *Wine Enthusiast*, and ninety-one points from *Wine Spectator*. I was fortunate to be able to taste both of these wines, as well as the 2009 FAY, and the differences between the three were eye-opening.

The winemaker for the 2009 vintage was Nicki Pruss, a soft-spoken brunette with baby-blue eyes and a passion for winemaking. She actually went to podiatry school and became a foot doctor but switched to wine when she was "bit by the wine bug," on a biking trip through the vineyards of France. A native Californian, she started her winemaker training at both Napa Valley College and UC Davis Extension and then answered an advertisement to be a harvest intern at Stag's Leap Wine Cellars in 1998. Nicki's strong science background and fascination for wine was appreciated by Warren Winiarksi, who was still working there at the time. He mentored her, and after working for seven years on the winemaking team, she was promoted to head winemaker in 2005.

"In Napa Valley in 2009, we had a cooler vintage than previous years," stated Nicki. "Therefore, we were able to let the grapes hang a bit longer on the vine to achieve more intense flavors and color. But we were fortunate in that we completed the harvest before the October rain came."

As I tasted the 2009 S.L.V. Cabernet Sauvignon, I detected clear signs of a cooler vintage in the glass. The wine was a deep red-black in color with a nose of forest floor and rich, dark fruit. On the palate the tannins were huge and not yet well integrated but would mellow out over time. There were notes of blackberry, coffee, and complex herbs and spice, and the wine had a long lingering finish. This was a massive wine that needed more time in the bottle or a juicy char-grilled steak to help tame the immense tannins. The oak seemed well integrated, and Nicki said she used 84 percent new French oak for twenty months. The alcohol was moderate at only 13.5 percent.

"The majority of the grapes in the bottle," said Nicki, "came from Blocks Two and Three, but at the wine's core is Block Four of S.L.V. where the oldest vines reside." As I sipped the wine, I thought of the huge shaggy vines, and it seemed fitting that they would produce such a powerful wine in a cooler year.

"How do you feel about S.L.V.?" I asked.

"I feel quite fortunate that S.L.V. has played an integral role in my development as a winemaker. In my opinion, the soil, the topography, the vines, and their interaction with the Stags Leap District mesoclimate creates an alchemy that is very special for the development of Cabernet Sauvignon. Based on where a block is located in the vineyard, there is enough diversity in the raw material between the blocks to create wines that produce what I like to call a 'painter's palette' of Cabernet Sauvignon."

"So it's a little bit like being an artist when blending the blocks?"

"Yes, there are nuances held within S.L.V. that allow the winemaker to create liquid art and 'paint' an exceptional wine: one with subtle power, restraint, and elegance. In my opinion, those words describe the 'fingerprint' of S.L.V. It is a vineyard that produces wines that can stand the test of time."

As I sipped the wine, I had to agree with her. This was a wine that was designed to age and had the grace and backbone to do so.

Next I tasted the 2009 FAY, which was more of a dark red-purple in color, with a fruiter nose of both red and black fruit with a hint of violets. On the palate, the fruity notes became clearer, expressing themselves as cassis and raspberry with toast and vanilla. Most distinctive, however, were the velvety tannins that glided across my palate, highlighting their clay soil origins. FAY's tannins did not attack my mouth, as did those of the young 2009 S.L.V. from volcanic soil. However, FAY's finish was warmer due to a slighter higher alcohol of 14 percent. The 2009 FAY was aged for nineteen months in 84 percent new French oak.

"You can definitely taste the difference between the two vineyards in the glass," Nicki commented. "S.L.V. is all about structure and subtle power, whereas FAY is approachable elegance and grace. They both have their places in the world, but often when we blend elements from the two vineyards together to create CASK 23, we achieve a higher level of synergy."

CASK 23 was actually started in 1974 when André Tchelistcheff was still the consulting winemaker. During that harvest André noticed that one lot of the wine from S.L.V. was so unique and compelling that he thought it should be bottled separately. Therefore, he put it in a French oak Oval that was labeled CASK 23. Over the years, wine from FAY was blended in, and today, CASK 23 is always a combination of both vineyards. The result is a harmonious mixture of the structure and power of S.L.V. married to the velvety richness and perfume of FAY.

As I lifted the 2009 CASK 23 to my nose, I was immediately overcome with the earthy perfume of violets. The wine was the same black-red color as S.L.V., but on the palate it exploded with rich black fruit, coffee, spice, and earthy notes. The level of complexity was quite high, and the finish seemed to go on forever. Here the tannins were well integrated with toasty oak, and the wine seemed perfectly balanced. Nicki said there was a bit more oak—90 percent—but it was also aged twenty months just like the S.L.V. The wine in my glass was clear proof that blending two great vineyards can result in something truly incredible.

Chapter Seven

Hirsch Vineyard

Fort Ross/Seaview AVA, Sonoma County

Climbing into my car to drive to Hirsch Vineyards, I noticed that the temperature gauge read a balmy 78°F degrees. It was a sunny day in May and I knew that it was about an hour and half drive from my home on Sonoma Mountain to the top of the coastal range overlooking the Pacific Ocean where Hirsch was perched. However, I was looking forward to the drive because I knew it passed through the quaint towns of Guerneville and Duncan Mills before arriving in Jenner on the coast.

As Old River Road wove its way through vineyards and towering groves of redwood trees, I caught my breath in awe several times at the beautiful patterns of sunlight filtering through the high boughs of the trees. It cast dappled shades on the bark, causing it to glow with red and orange light, which reminded me of stained glass windows in a cathedral.

On my left the Russian River snaked through the redwoods, growing larger as it made its way toward the Pacific Ocean. I caught glimpses of brightly colored kayaks on the water, with paddles flashing in the sun. Suddenly I turned a corner, and saw a huge cloud of fog swirling over the mountains on the right. It drifted in a soft white web, settling over a hillside of dark green redwood trees like a nourishing blanket of moisture.

Glancing at the car thermometer I was not surprised to see that the temperature had plunged to 56°F degrees, even though I had traveled only 30 miles. The Sonoma Coast was generally at least twenty degrees cooler than the inland valleys of the county, and when the fog was rolling in, it could be even colder. This reminded me of my love/hate relationship with fog. Though I knew its chilling influence provided moisture and protection for both redwoods and vineyards, there were days when it didn't lift until noon, making for many gray mornings.

When I reached the hillside town of Jenner, its buildings were barely visible, wrapped in thick strands of fog. As I rounded the last curve in Jenner, heading north on Pacific Coast Highway 1, I glanced out the window to see the stunning sight of the Russian River meeting the ocean. Here gray sand bars funneled the wide mouth of the river into a smaller channel, until it merged into the salty water of the Pacific with a frothy bounce of choppy waves. Large craggy rock formations rose in the ocean a slight distance from the shore, and I could just make out the silver bodies of the harbor seals beached on the sand.

The road twisted and turned, climbing higher, and each turn brought a more terrifying plunge to the ocean as the sheer rocky cliffs increased in height. This is why they call this stretch of highway the Extreme Sonoma Coast, I thought. That, and the fact that pioneers like David Hirsch were crazy enough to plant vineyards on the coastal mountains of this area, creating an "extreme" location for viticulture.

Suddenly as the road continued to climb even higher, I rounded a corner and came into the sunlight. The swirling fog dropped away, and I could see cows grazing on the steep green hillsides and black California vultures soaring in the air currents. Looking at the cows, I was amazed they didn't slip down the steep hill into the ocean, but they continued to eat calmly, completely unaffected by their million dollar views of the Pacific.

Glancing down the cliffs, I saw the fog was pooled below me, covering the sea and everything else in a thick white lake of dense moisture. Then a sign appeared on my right for the Meyer's Grade turnoff, and I knew from my map that this was the direction towards Hirsch Vineyards. Heading up the mountain, away from the ocean, the air gradually became warmer and my temperature gauge showed 68F.

Wildflowers in hues of yellow and orange dotted the green fields, and small vineyards appeared occasionally along the narrow twisting road that was filled with potholes and patched asphalt.

Eventually the paved road ended completely, and I found myself driving along a dusty dirt track with no other vehicles in sight. After driving almost twenty minutes, I encountered a flock of sheep eating the grass alongside a vineyard. They were large white wooly sheep, and I had flashback to vineyard visits in New Zealand where sheep are allowed to graze amongst the vines.

Sheep Near Vineyards on Drive to Hirsch

Finally, after another ten minutes, a cluster of buildings appeared on my right and I saw several houses, a shed, and stacks of grape picking bins piled high outside a large barn. There was no sign that announced Hirsch Vineyards, but since the road dead-ended in front of the barn, I knew I must have arrived.

A large black lab ambled over to greet me as I opened the car door, and when I gently extended my hands to let her smell me, she licked them instead. Patting her on the head, I tried to wipe off some of the dog slobber on her fur.

"So you found us."

A cheerful voice hovered over my head, and I looked up to see a slender young woman dressed in faded blue jeans, a pale gray sleeveless t-shirt, and orange flip flops standing in the doorway. Her brownish blonde hair was pulled back in a tight bun that emphasized blazing blue cornflower eyes in a pretty face devoid of make-up. I knew she must be Jasmine Hirsch, daughter of David Hirsch, and current marketing and public relations manager for the estate.

Jasmine reached forward to shake my hand, and remembering the dog slobber, I quickly rubbed my palm on my pants. She laughed. "I see you've already met Ruby, our vineyard dog. Don't worry," and she grasped my hand in a firm confident handshake and then motioned me inside the barn-like building. As we entered I could see it was a combination office and wine laboratory filled with desks, computers, filing cabinets, lab equipment, sinks, and a round table with a large vase of purple tulips.

"Welcome to Hirsch Vineyards," Jasmine said. "As you can see we are rather a casual operation." She gestured to the room. "Dad's out in the vineyard somewhere, but before we go find him, would you like to hear how he discovered this property?"

History of Hirsch Vineyard

Born in New York of Romanian and Polish parents, David Hirsch grew up in the Bronx and studied literature at Columbia University before dropping out of college to move to California in the 1960s. He ended up in Santa Cruz where he met lifelong friend, Jim Beauregard, who had planted a small vineyard in the Santa Cruz Mountains. Jim's enthusiasm for viticulture was infectious and soon David became fascinated with vineyards and wine.

While keeping his day job as owner of an import business, David began searching for vineyard property. He considered parts of Santa Cruz, but was concerned it was getting too crowded. Therefore he began to look further north, armed with the clear knowledge that he wanted to be a in remote location but close to the ocean. Eventually in 1978, he saw an advertisement in the *San Francisco Chronicle* for land located on top of the coastal mountains of the Sonoma Coast.

"It was an old sheep ranch," explained Jasmine. "Dad came up here and fell in love with the place, but the land had been destroyed by over grazing, and most of the redwoods had been cut down. However, he wanted to bring it back to life, and needed a cash crop. He knew the Bohans down the road had put in the first vineyard up here and were selling the grapes for bulk wine, therefore Dad decided to plant a vineyard to produce premium grapes and to help heal the land."

Hirsch Vineyards With Redwoods and Pacific Ocean

After purchasing the property in 1978, David Hirsch planted one acre of Pinot Noir and two acres of Riesling in 1980, in what is today known as the Old Vineyard. Situated at 1500 feet high on the Pacific Coast range, the ocean is visible from the vineyard on clear days. After several years when the vineyard came into production, David began selling the grapes to Sonoma County winemakers.

By 1987 David was able to quit his day job and move to Hirsch Vineyards with his family. In 1989 they grafted the Riesling over to

149

Pinot Noir, which was a more profitable crop. Then over the years they planted more vines, including 44 acres of Pinot Noir and Chardonnay between 1990 and 1996, and an additional 25 acres in 2002 and 2003. They also converted the vineyard from conventional farming to biodynamic principles.

By 2000 the high quality of the grapes produced from Hirsch Vineyards were receiving much positive attention from winemakers, and David was able to obtain more money from his crop. Eventually he decided that, due to the growing interest in wines made from Hirsch grapes and because he wanted more feedback on grape growing, he would begin making wine.

Through a combination of self-study and hiring consultants David was able to produce their first vintage in 2002. Since then Hirsch wines have grown in acclaim, and today Ross Cobb is the full-time winemaker. His experience working with Hirsch grapes at Williams Selyem and Flowers made him an ideal candidate for the job.

David continues to devote himself to the vineyard, which now has the distinction of being the oldest premium grape vineyard on the true Sonoma Coast. Over the years, David has been a huge proponent of creating a specific AVA for the region, and in 2012 won this battle. Today Hirsch Vineyards is part of the Fort Ross/Seaview AVA.

Touring Hirsch Vineyard

"Come on. I'll take you on a tour of the Old Vineyard."

Jasmine put on a pair of dark brown sunglasses and motioned me towards the door. Ruby followed us as we headed across the dirt parking lot to a small block of vineyard that had been newly replanted. Mounds of brown soil were heaped around small sticks of rootstock, and a colorful cover crop of green grasses, red clover and bright blue flowers ran down the rows between the mounds.

"But I thought you said this was the Old Vineyard," I said with disappointment, wondering what happened to the vines planted in 1980.

Jasmine glanced back at me as she walked into the block. "This is the part of the Old Vineyard that we are replanting, but there is still a large section of the original vines over there."

She pointed into the distance, and I could see some gnarled old vines just beyond the edge of the block. They march solidly up a hill to where a stand of redwood trees formed a perimeter to the vineyard.

Jasmine stopped in the middle of the row, and Ruby rubbed her black coat against Jasmine's leg. "I have to admit I was shocked at first when Dad said we should pull out some of the old vines, but you see he has a 200 year plan for the land. He is an avid reader, and has read most of Rudolph Steiner's books on biodynamics, so he has a long-term vision for the ranch."

"An avid reader," I repeated. "That's right, he was a literature major at Columbia."

"Yes, and one of his other favorite authors is James Joyce. He likes to read very complex prose, and he tries to get me to read it too." Jasmine flashed a quick grin, and then she bent down to look more closely at one of the dirt mounds with a small stick in the center.

"We are replanting this because it was not performing as well," she said. "So we took budwood from the original vines and are keeping it in our greenhouse. In one year we will graft over."

"So the budwood is the famous Hirsch Pommard clone?" I asked remembering the information from the website.

Jasmine nodded. "Yes, we think it was originally from Oregon, and is a combination of Pommard and Wädenswil. Over the years it has adapted well here, so some people refer to it as the Hirsch Pommard clone. Let's go look at the old vines."

When we reached the older section of the vineyard, Jasmine stopped and gestured towards the horizon in the distance. Glancing up I was amazed to see the vast expanse of the Pacific Ocean with a blanket of white fog hovering over it. The vineyard sloped slightly downhill towards the ocean, and I noticed another large group of redwood trees at the border of the vines. In the sky above, a red tail hawked circled slowly, enjoying the ocean breeze along its wingspan.

"What a view!" I exclaimed. Standing there on the edge of the mountain, with the vineyard, redwoods, and ocean in the distance, I felt a huge sense of euphoria and peace. It was very uplifting. "I can see why David put the vineyard here."

Vines in the Old Vineyard Block

Jasmine nodded. "Yes, I never get tired of this sight."

It was a few minutes before I could drag my eyes away from the magnificent view to focus on the Pinot Noir vines in front of me. They were trained on VSP trellis with spur-pruned cordon. Thick wide trunks with dark shaggy bark proclaimed their age. Bright green leaves unfurled from the shoots, and I could tell from their size that they were almost completely formed for the season. The small grape clusters were also visible as tiny round balls, but it was obvious they had not yet progressed to the flowering stage yet.

Looking down the rows, I noticed that some vines had truncated cordons, and in a few places new shoots were growing from vine trunks that had been sawed off about six inches above the ground.

"This is the Mother Block," announced Jasmine, "and as you can see we've had some Eutypa issues where we've had to cut off part of the cordon, or in some cases, cut the trunk all the way back to grow a new shoot."

"Is it because of the rainy climate here?"

"Yes. Keep in mind this used to be a redwood forest before the land was cleared for sheep grazing. Because of this we still get an average of 80 inches of rain per year. Therefore having too much moisture during pruning can cause Eutypa, which kills that part of the vine."

I nodded, realizing that the very wet climate of a redwood forest could have some negative implications for grape vines.

"And the fog," I asked gesturing to the blanket of fog on the ocean, "how often does it come up this high?"

Jasmine smiled. "Oh, the fog," she said. "One winemaker I know calls it 'refrigerated sunshine.' Since we are at such a high elevation we are not enveloped in the fog as often as other parts of the Sonoma Coast or Russian River AVAs, but we do often have a layer in the early mornings and most evenings. As you know, the cooling influence of the fog allows the grapes to maintain their acidity and preserves the flavors."

She reached out and touched one of the tiny clusters of newborn grapes, before continuing. "Dad says our vineyard on the top of this ridge has the perfect of amount of light and heat. We have the sunshine when the fog burns off, but since we are on top of the mountains and close to the ocean, the sun is not too hot. This allows our grapes to ripen perfectly, so you can taste the sunshine in our wine but not the heat. Shall we go check out a newer block?"

I looked out once more over the Old Vineyard, realizing that these were the oldest premium Pinot Noir vines on the Sonoma Coast. The

Pinot Noir grape is known for its delicate skin and difficulty in growing. It throws temper tantrums if it is too hot, resulting in stewed overripe fruit, and produces thin green tasting wine if the climate is too cool. There are only certain locations in the world where high quality Pinot Noir can be produced, and the cooler climate that encompasses both the Sonoma Coast and Russian River is one of those rare places on earth.

Ruby brushed my leg as she galloped after Jasmine who was already several steps ahead of me, walking back the way we had come. The brightly colored cover crop with its red clover and blue flowers tickled their legs.

Suddenly Jasmine stopped and bent down. "This is one of my favorite flowers," she called out.

Approaching her, I looked down at the ground to see what she was pointing at and noticed she wore bright pink toenail polish with her orange flip flops. Her fingers were caressing one of the delicate blue flowers.

"What is it called?" I asked.

"I think it is phacelia," responded Jasmine. "I love the blue color, and I think it is supposed to attract honey bees and other beneficial insects."

"Beautiful."

As we continued walking Jasmine shared more information about her background. She grew up on the property, and remembered when they had to use gas lamps before installing their own electric line. After high school, she went to college on the East Coast and then traveled around Europe for a while. It was when she was working in New York that a friend suggested she come back to the ranch and help her Dad.

"He said that my Dad was doing something special here restoring the land and converting to biodynamic farming," said Jasmine. "He suggested that with my marketing background that I could help with that."

"And do you like living out here on the remote Sonoma Coast?"

"Yes, I love being back here on the land, but I also travel a lot in my job. This ranch, with its vineyards, is the center of gravity for our family.

My sister comes to visit and her kids love driving around in the tractor with my Dad. Having this place to come home to is wonderful."

She paused and glanced over at me. "You remember I mentioned my Dad has a 200 year plan? He wants to heal the land and keep it in our family. I didn't understand what that meant until I went to Europe where it is common to have one family farming the same land for centuries. That concept doesn't really exist in the US, but we want to make it happen here."

"Admirable."

"And I enjoy thinking about the history of this place," she continued. "Just yesterday I found an arrowhead. It was a reddish blonde color, perfectly formed with notched edges and about this long." She held up two fingers to show a length of about one inch. "We've discovered the Native Americans passed through this area because there are several natural springs and a small midden, which is one of their ancient trash piles where they threw oyster shells and bones."

I nodded thinking about the three major Native American tribes that lived in Sonoma County; the Coastal Miwoks, Pomo, and Wappo. It was inspiring to realize they had roamed these ridges with views of the ocean. Today the descendants of some of these tribes were managing very successful casino operations in Rohnert Park and the Alexander Valley.

We had passed the barn and were walking up the hill to another vineyard block. As we reached the top of the hill, I could see rows and rows of vines flowing over the undulating hills and into the distance.

"Wow."

"Yes, it's a wonderful site," said Jasmine. "This is Block 15 right in front of us, which is a new Pinot Noir block. Off in the distance you can see blocks 6, 12, 7, 8, and 11. Several of them are Chardonnay."

I followed Jasmine's pointed finger out into the distance where I could see more blocks of vines. They spread out over the hills and valleys, marching up and down slopes. It looked like someone had thrown a beautiful patchwork quilt in different shades of green over the land.

"Since you're in charge of marketing," I half joked, "why haven't you developed more creative names for the blocks than just numbers?"

Jasmine grimaced and turned to look at me. The sun reflected off her taupe sunglasses, as she replied with a wry smile. "Dad doesn't like fanciful names. When I suggested we come up with more creative names he told me, 'Go ask the site. It will tell you what it should be called. Do not impose your ideas on the site.'"

I laughed. "Well that is one way of naming the blocks."

"Looks like you can ask him yourself," said Jasmine. "Here he comes." She nodded her head to indicate David Hirsh was behind me.

Turning around I saw a slender man of medium height walking up the hill with energetic purposeful strides. He wore faded blue jeans, a short-sleeved gray shirt, and a straw hat over curly gray hair. As he got closer I saw he had a full white mustache, tanned skin from spending many hours outdoors, and black square-cut sunglasses to shield his eyes.

After introductions, Jasmine excused herself to go back to office and David turned to me. "Let's go check out the vines," he said, and I detected a faint hint of a New York accent in his voice. We walked into the rows of Block 15, and I could see they were younger vines on VSP trellis with tighter spacing than the Old Vineyard. The trunks of the vines were thinner, and the cane-pruned shoots seemed more delicate as they draped themselves along the wire, their healthy green leaves sharply highlighted against the bright blue of the sky.

"When was this block planted?" I asked.

"2002. It's Pinot Noir with a combination of Pommard, 777 and Swan clone." He reached out to tuck a wayward shoot under a wire.

"This is a very impressive site. What caused you to decide to buy this land in the first place when very few people were living out here?

"I didn't buy this land. It bought me." His voice was forceful and passionate. "When I came here in 1978 the top soil was gone, the land was bare. It had been over-grazed by sheep." He gestured his hand towards the vineyard. "I wanted to bring it back to life, so I needed a cash crop. I planted grapes to make some money to help heal the land. I didn't intend to start a winery." He turned to stare at me, and the light flashed off his dark sunglasses. "But winemaking is the last step of farming. The wines give us feedback on what we are doing right or need to modify in our farming."

David Hirsch

"So tell me about your philosophy of biodynamics."

David's gaze moved out across the vineyard. "In the beginning we needed more control, so we used conventional viticulture, but over time I came to understand more about organic and biodynamic farming. I read Steiner's books on biodynamics, called *Spiritual Foundations for the*

Renewal of Agriculture. In addition I asked Ted Lemon to give us a seminar on how to farm a vineyard using biodynamic techniques."

"So has it paid off in your estimation?"

"I believe so, though we've only been using biodynamics for the past few years." He paused, and when he spoke again his voice was lower. "You know, there is an underground world and an above ground world for the vineyard. Biodynamics is all about nurturing the soil back to health."

I nodded, remembering the basic tenants of biodynamics with an emphasis on returning the soil and environment to its original state before man intervened. It is a step beyond organic farming, in that there is a spiritual component to it. Practices such as following the phases of the moon, burying cow manure in horns, and applying preps to the soil and leaves of the plant, are examples of the more philosophical side of biodynamics. Those who follow the tenants are usually very passionate about the benefits, though there are naysayers who have termed it "voodoo viticulture." It was obvious that David Hirsch was on the side of the proponents.

"I try to farm each vine individually," David was saying. "I enjoy spending time with the grapes. It is more like a dream life, as the Aborigines of Australia refer to dreaming."

As I listened to his words, I was reminded of his strong interest in literature. It was obvious from his statements that he was very well read and a deep thinker – more like a vineyard philosopher with strong opinions on how the earth and man should interact.

David reach out and touched one of the green leaves of a Pinot Noir vine. "You know plants don't have an interior emotional system," he said, gently stroking the leave. "They are dependent on their environmental. There is not such thing as a sick plant, just a dysfunctional environment. We need to make the environment around the plant more healthy."

"That's a commendable philosophy.

"It's the same with people," he said, staring at me from behind the black sunglasses. "We eat too much junk food now. That's why I hired our chef and gardener, Ryan. He planted an organic garden and raises chickens for eggs. We try to eat as healthy as possible here, and we try to do the same for the vines."

Hirsch Vineyard Blocks Flowing Over Hilltops

I reached down and picked up a hand of soil. It was a light grey brown and trickled gently through my fingers. "Tell me more about your soil."

"We live on the San Andreas fault here, and our soil is very complex. It is called the Boomer Mendocino series, and is a combination of clay loam, sandstone, shale and various types of rock. We follow the biodynamic practices of using 500 prep on the soil and 501 prep as an aerial spray. This nurtures the vineyard and helps bring the environment back into balance."

"I've heard that biodynamics requires a lot more work than conventional viticulture?"

"Yes," said David, "but we have an excellent crew of employees here who tend the vineyards. Everardo Robledo is our Vineyard Manager. He helped me graft the Riesling to Pinot Noir in 1988, and then decided to stay on the property. He lives in one of the houses here with his wife and four children, and is one of the smartest viticulturist I've ever met." David's voice was filled with pride and respect when he mentioned Everardo.

"So he's a member of the famous Robledo Family that are experts in California viticulture?" I asked, referring to a group of brothers who had immigrated from Mexico to work in Napa and Sonoma vineyards as master grafters. They were known as some of the best viticulture experts in the state.

"Yes," David said proudly, "and he helped plant the more than 90,000 vines we have on the ranch now. He knows the vineyard instinctively and responds to what is happening." David pauses for a minute before continuing. "You know, I believe there is a strong split in knowledge between the daily vineyard worker who knows the land so well and the UC Davis grad who thinks he knows the land. Everardo is the true expert."

David's opinion about the expertise of the vineyard workers from Mexico is one that is echoed by others. In fact some experts suggest that the strong work ethic and almost mystical connection the Mexican worker has to the land is one of the major competitive advantages of the California wine industry. Indeed, it is similar to Rudolph Steiner's thoughts about the peasants of Europe:

> *"I have always considered what the peasants and farmers thought about their things far wiser than what the scientists were thinking. ... I have always been glad when I could listen to such things, for I have always found them extremely wise, while, as to science in its practical effects and conduct I have found it very stupid (p. 3). "*

"For example," David was saying, "In 1998, which was an El Nino Year, we had 150 inches of rain. It rained so hard the roots started sticking out of the soil. You see that block over there," he pointed towards the Eastern hills where a block of vines marched solidly up a very steep slope. "That is one of our best Chardonnay blocks, and it was almost destroyed. Everardo and I worked for days replacing the soil to save those vines. He knew instinctively what to do."

The 1998 El Nino was legendary for its torrential rainfall, a hallmark of an El Nino year where the weather patterns on the Pacific bring in much more rain and cooler weather than usual. I could easily imagine David, Everardo and other workers battling the elements on that steep slope in the distance to save the vines.

We began walking back down the hill towards the winery, and David spoke of books he had read and some of his favorite essayists. As we approached, a well-muscled man wearing blue jeans, work boots, and a long sleeved charcoal gray shirt came walking towards us from the direction of the Old Vineyard.

"Everardo," David called out to him.

A broad grin split Everardo's face, and he quickened his step to shake hands with David before turning to me. He gripped my hand in a gentle handshake, and his brown eyes were warm and friendly in a tanned face. He wore a navy blue baseball cap pulled low over his forehead, concealing a full head of short-cropped black hair.

"Everardo will tell you everything you need to know about the vineyard," said David. "He knows the logic behind every clone, rootstock, and spacing decision."

Everardo beamed his thanks to David in a broad smile, and then beckoned for me to proceed him along a small grassy path that skirted a line of redwoods. Ruby appeared from out of the bushes and led the way, her black coat shining in the sun. As we walked I realized we were headed back toward the Old Vineyard but on a lower circuit that ended further down the hill in front of a block of newer vines.

"This is Block 16," Everardo said, with a charming Mexican accent spicing up his words. "We planted this in 2003. It includes three clones of Pinot Noir – Pommard, Swan and 114."

Everardo Robledo, Hirsch Vineyard Manager

The vineyard seemed elegant with thinner trunks and cane-pruned shoots spread along a VSP trellis of about 4-foot in height. The row spacing was a generous 8 feet wide, but only 3 feet between the vines.

"So why is the Old Vineyard on cordon and the newer blocks cane-pruned?" I asked.

"Good question," Everardo said. "In the beginning they trained the vines as cordon, but over the years we realized that cane-pruning works better here because you have less problems with Eutypa. You know we have lots of rain here?"

"Yes."

"So cane pruning does not require that we make such large cuts in the wood. Therefore there is less risk of Eutypa disease."

"Can you show me how you prune these vines?"

"Sure." Everardo placed his hand upon the upper trunk of a vine where this year's canes were neatly tied to the wire, each going in the opposite direction for about 1.5 feet. "So in January we cut off all of last year's shoots except for these two." He pointed to the two canes tied along the wire, and then gently touched the small shoots with their bright green leaves and baby grape clusters climbing upward from the cane. "We leave 6 buds per shoot for a total of 12 buds. Each bud will produce one shoot, and each shoot produces leaves and two clusters of grapes."

As he talked I noticed how large his hands were upon the vines, and how tenderly he touched them. As he cupped a new cluster of baby grapes in his palm, I could see the calluses and dirt on his hands. They seemed so capable and filled with knowledge and love for the vines.

"So how long have you been working in vineyards?"

Everardo straightened up from bending over the vines. "I came to California when I was 14 from Michoacán, Mexico." He smiled as if lost in thoughts of a faraway time. "My mother, she begged me not to be a vineyard worker. She wanted me to go to school and become an architect." He shrugged his large shoulders and smiled again. "But my father and uncles got me a job in Napa and taught me the art of grafting vines. So you see, we have all become master grafters."

"What you do is a different type of art," I said. "Perhaps you should tell your mother you are an architect of vineyards."

Vineyard Specifics for Hirsch

As we continued walking through various blocks of Hirsch Vineyards, Everardo shared more specific information about the property. They have a total of 72 *acres* under vine; with 60 blocks to differentiate the various soil types, exposure and topography. Grape varietals are 95% Pinot Noir and 5% Chardonnay.

According to their *soil* report, there is a wide diversity of different soil types ranging from Rocky, Sandy and Clay Loams, Volcanic, Plastic Clays, and Reddish Loams. The *elevation* of the Old Vineyard is 1500 feet above sea level, and the other blocks range from 1400 to 1550.

"You can see the ocean from other parts of the vineyard too," Everardo explained.

Summer *temperatures* average 85°F for a high and 50°F as a low at night. Fog rolls in most evenings in summer and burns off by mid morning. *Rainfall* is usually around 80 inches per year.

Vineyard Specifics for Hirsch

Total Vineyard Acres	72 acres, 60 blocks
Varietals	Pinot Noir and Chardonnay
Soil	Rocky, sandy and clay loams, volcanic, plastic clay & reddish loam
Elevation	1500 feet above sea level
Average Temperature	Summer averages 85°F for a high and 50°F as a low. Rainfall averages 80 inches per year
Rootstocks	Self-Rooted, 3309, 101-14, AXR, SO4, Riparia, 4453, SC
Clones	Pommard (Hirsch's), Mt. Eden, Swan, Calera, 114 and 777
Sun Exposure	Mixed, but primarily south to west row orientation
Spacing	Older vines on 9 x 5, 10 x 5, 10 x 2.5, and 12 x 6. Newer vines on 8 x 5, 8 x 3.5, 8 x 3.25, and 6 x 3.25
Trellis Systems	VSP

"My biggest concern is the rain," Everardo said. "We have extreme weather here with too much rain some years." He shook his head as if remembering past vintages of torrential rain, and I remembered David describing how they had to work very hard to save the vines during El Nino years.

"So what do you do about it?"

Everardo shook his head and shrugged his shoulders. "We can't control it." He looked up at the sky. "Today is sunny, so I don't worry. I like the vineyard when it is like it is today - healthy and in the sun." A large smile lit up his face and his teeth flashed white, before a frown eclipsed the smile. "But other times I worry so much about the weather and what it will do to the vines."

I nodded, recognizing the deep emotions and sense of care and responsibility he felt towards the land and vineyard. As we continued walking, he described the philosophy behind their selection of rootstock and clones.

Rootstock for the Old Vineyard is "self-rooted," but rootstocks for the newer blocks were determined based on the soil type. The most common are 3309 and 101-14, but in some areas SO4, 4453, SC, Riparia and AXR are used.

"Because the soil in some blocks has more clay and others have more sand," explained Everardo, "we picked the rootstock that worked best with the soil. You know, the clay soils hold water more, but if it is sandy, then the water drains quickly. The soil changes a lot here. That is why we ended up with so many blocks."

Clones include Pommard (also called Hirsch's Pommard), which is the most common in the vineyard, as well as Mt. Eden. In the later plantings, Swan, Calera, 114, and 777 were added in different blocks. *Sun exposure* is mixed, but primarily with a south to west row orientation, and all *trellising* is VSP.

"David said you helped with the grafting of the original Riesling vines to Mt. Eden Pinot Noir," I said.

Everardo nodded modestly. "Yes, in 1988 when I first came here."

"I notice that the spacing is different in some of these blocks," I commented, swinging around to look at the Old Vineyard as we slowly walked back toward the barn.

"Yes, the Old Vineyard was planted at 9 x 5 feet, but we have changed the spacing over the years, depending on the soil and slope. Now we are mainly using 8 x 3 on the newer blocks, but there are many different sizes because the land changes."

I nodded, thinking of the vineyard data sheets provided on the website that showed *spacing* ranging from 12 x 6 in the Chardonnay blocks, to 10 x 6, 10 x 5, and 10 x 2.5 in older Pinot Noir plantings, and 8 x 3, 8 x 3.5, 8 x 5, 7 x 3.25 and 6 x 3.25 in some of the newer blocks. According to the website:

> *"A cursory review of the vineyard data shows the large difference in planting density between the old and new fields, (and) ... an expansion in the kind of rootstocks used. New methods of site mapping, investigation, and soil analysis were employed. Given our environment, clones are secondary, We have sought clones from older, well-farmed sites that are untainted by the contemporary processes of genetic selection and industrial nursery propagation: in looking for scion wood with soul, we will take a little blemish."*

Farming Methods at Hirsch - Biodyanmics

"Can you describe your farming process?"

We were walking through the last row of the vineyard and Everardo stopped to reach out and pull a wayward sucker off a vine near to us. "Well, we usually start pruning in January and the canes are cut up and put in the compost piles. Since we are using biodynamic methods, we make our own compost and Ryan creates the preps. I think he is making nettle tea today."

I remembered David describing Ryan's role as being chef and organic gardener, which included the responsibility of growing and preparing the plants for the biodynamic preps and fertilizers. Though using biodynamic farming methods, Hirsch Vineyard was not seeking *certification* with Demeter, the international organization that verified biodynamic practices.

"We *fertilize* twice a year," Everardo continued. "We put down compost in the middle of the rows before harvest, and then below the vines in spring. Cover crops are planted before harvest with the compost, and later tilled and mixed into the soil."

"The cover crop is beautiful," I said gesturing to the brightly colored flowers intermixed with the green. "What does it include?"

"A combination of clover, oats, beans, and flowers. We have a lot of nitrogen in our soil here that can make the vines unbalanced. So we are trying to balance them naturally with our cover crop. I've discovered that oats, daikon radish and cauliflower can help reduce nitrogen."

"Fascinating. What about the biodynamic preps?" I was curious to know how they were applying Rudolph Steiner's nine preparations labeled 500 to 508. The first two, preps 500 and 501, are the most common and used for stimulating humus or organic matter in the soil. They are mixed with water and sprayed on the soil or in the air above the vines. The other preps are used primarily in compost. All are made of natural materials such as quartz, nettle, chamomile and other ingredients.

"We use both the 500 and 501 biodynamic preps," replied Everardo. "With the air prep, we spray it over the vines using our small ATV tractor. We also do this to spray sulfur to protect from powdery mildew."

He explained that for *weed control*, they mowed under the vines. *Canopy management* is done 100% by hand. "We sucker and shoot thin in April and May. After bloom we will drop excess clusters, and then do a second pass at verasion to drop shoulders if necessary."

The main *disease control* issues are Eutypa and powdery mildew, which are both exacerbated by the high rainfall average in the region. "Beginning with budbreak in March, we begin spraying with sulfur every 15 days," explained Everardo, "then we switch to organic Kaligreen at bloom."

"Kaligreen," I said surprised. "Isn't that only used when you already have powdery mildew?"

"We also use it as a preventative because of the rainfall here. It's 80% baking soda anyway, so it is quite natural. To reduce Eutypa, I already explained that we are switching to cane pruning."

Farming Practices at Hirsch Vineyards

Certifications	Not certified, but using biodynamic practices
Fertilization	Cover crop, compost, and biodynamic preps
Weed Control	Mowing
Canopy Management	100% suckered and pruned by hand
Disease Control	Sulfur and organic fungicides to prevent powdery mildew. Cane pruning to reduce Eutypa risk.
Pest Control	No major issues, using owl boxes and beneficial insects
Irrigation	Drip irrigation
Technology	Petiole analysis
Harvest Measurements	By tasting and low brix – usually 22 to 23

Pest control is not a large problem at Hirsch Vineyards. They are using owl boxes and beneficial insects to patrol for gophers, small birds, and destructive bugs, however some birds are still an issue.

"We have some turkeys that eat the grapes," reported Everardo, "and we must net near the trees to protect from small birds that live in the trees and will eat the grapes nearest to them. We also have snakes, but they are not a problem. We just leave them alone."

Drip irrigation has been established in the vineyard, but is not always necessary to use due to the high rainfall. Very little *technology* is used, with the exception of petiole analysis every few years.

"We don't use much technology here," said Everardo. "There are no weather stations or frost alarms. We do some petiole analysis, but not annually." He paused and looked at me with a shy smile. "David and I are the technology. The vines tell us what they need."

For *harvest measurements*, Everardo explained that David and Ross, the winemaker, came to the vineyard to taste the grapes for several weeks in advance. "They generally pick at 22 or 23 brix," he reported.

"That low?"

Everardo nodded. "Yes, David likes low alcohol Pinot Noirs, and of course, we harvest all the grapes by hand."

Economics of the Vineyard

The *average yield* for Hirsch Vineyards is 2.5 tons per acre, resulting in 180 average tons per year based on 72 acres. *Farming costs* range from $6500 - $7500 per acre excluding depreciation.

"We have 15 full-time employees," explained Everardo, "and will add five more part-time people during harvest. The hourly rate is $11, plus we provide worker housing and insurance."

"Do you pay by the ton during harvest?"

Everardo nodded. "We pay by the ton, and harvest at night from 2am to 7am. The workers can pick 12-15 tons during that time, so they can make a lot of money."

Economic Viability of Hirsch Vineyard

Average Yield	2.5 tons per acre
Total Average Tons Per Year	180
Costs Per Acre	$6500 - $7500
Revenues	$14,000 - $16,00 per acre
Economic Health	Good

In terms of *revenues*, Hirsch Vineyards is still selling 2/5th of its Pinot Noir crop and retaining the remainder for their own wine label. "In 1994," David reported, "I was selling my grapes to Kistler, Literai, Siduir and Williams Selyem. In 2000, I was getting $1200 per ton. Now we sell by the acre." Revenues per acre are calculated based on individual contract, and can range from $14,000 - $16,000.

Total wine production for Hirsch Vineyards averages 5500 cases per year, with bottle prices ranging from $34 for the estate Pinot Noir and $55 for the Chardonnay to $60 for the signature San Andreas Fault Pinot Noir and up to $85 for reserve wines. According to David the economic health of the vineyard is "good."

Soul of the Vineyard – "Healing"

To understand the "soul" and purpose of the vineyard, I consulted its founder, David, and asked him to describe it with one word.

"One word," he repeated, and then responded immediately without a need to think about the answer. "Healing," he said firmly. "My job is to heal this place."

We were standing on the hilltop near Block 15, and as far as the eye could see healthy green rows of vines flowed along the swells of the land, echoing the movement of the ocean waves far below. Occasional stands of tall redwood trees provided an accent as they rose tall and proud along the crests of hills or nestled in small hollows, their dark green fir in contrast to the bright green vines.

"When I came here this was a devastated piece of land," he continued, his voice confident and strong, "and I needed a cash crop to get money for the restoration. Now we've come full circle and the land is healing, but it will take 200 years."

"200 years?"

"Yes, that is the amount of time we really need to return to the land to its original state."

"Can you describe the best part of working here?"

"Walking the land everyday." His gaze moved away to hover over the panorama of the vines flowing over the land, and his voice was filled with emotion when he continued. "Living and breathing in this magnificent place."

There was a pause in the conversation, and then he headed back down the hill towards the barn now converted to a winery. Following him, I struggled to ask my next question, not prepared for the answers that spewed forth in a diatribe.

"What is difficult about living and working here?"

He stopped and swung around to face me. "The worst part is living in this county with all its crazy regulations. They say they want to preserve the rural country, but the zoning rules are destroying the small family farmer. It is extremely difficult to make a living off of the land with all of the rules. For example, you can't build another house unless you have more than 240 acres. Well, how can my daughters and son live here and each have their own house? It's not possible."

"Yes," I started to say.

"Another example is worker housing," he continued. "Do you know how long it took me to get permission to bring in those trailer houses for the workers? A very long time, and during my battle to get permission to do so, I discovered that some of the vineyard workers I was hiring in Graton were actually camping in the woods because they had nowhere else to live. In fact, one worker I picked up asked if we could stop on the way back here so he could collect his sleeping bag and other stuff he had left at his campsite. And here I was fighting with the country to get housing here for the workers."

His voice was filled with passion as he spoke, and he waved his hands and arms in emphasis. "Now, at long last, I have housing for them," he finished.

"That's wonderful that you achieved this," I said in admiration, knowing how complicated the issue was of worker housing for seasonal immigrants. The talented migrant workers from Mexico with their strong work ethic were considered by many to be the backbone of California agriculture, but government regulations and politics often made it difficult to insure they had appropriate housing. It was a long running battle, but it appeared that David Hirsch had at least won one of the skirmishes.

Ruby Near New Vines

We walked down the hill together, a tentative silence between us. As we approached the old barn that David had turned into a winery, he pointed to a tall redwood tree on the edge of the dirt driveway nearby.

"Do you see that redwood? I remember when I first brought my family here and the tree was much shorter. I have a photo of my daughter when she was young sitting on a horse next to that tree." His voice was calmer now and filled with warmth from the memory. "It is amazing how much taller the redwood is now."

"So what have you learned from the vineyard over the years?"

"Humility," David said quietly. "The vineyard has taught me so much over the years, and I've grown to recognize how little we know." He paused, and then continued, "We truly don't know what we don't know."

As I climbed into my car later, Ruby ambled over to say goodbye. This time she didn't lick my hands, but offered her back for me to pat. Running my hands along her glossy black fur, I could feel the smooth muscles underneath and the warmth that signaled she had been resting in the spring sunshine. Her gentle goodbye was much appreciated.

Driving slowly away over the dirt road I passed the Old Vineyard with its rows of vines stretching down the hill towards the uplifting vista of the redwoods and ocean in the distance. But now the layer of fog had lifted and I could see the hazy blue water fading into the horizon. I stopped the car in awe, and felt the same sense of uplifting peace descend on me that I had felt earlier when viewing this site.

As I sat there savoring the feeling, David's words of "healing and humility" fluttered through my mind, and I realized how fitting they were. The breath-taking beauty of the view inspired not only a great sense of humbleness and gratitude to witness, but imparted a healing sense of peace. And I wondered if, in his efforts to heal the land, David had created an environment that could heal others.

Signature Wine – Hirsch San Andreas Fault Pinot Noir

With 95% of the vines at Hirsch Vineyard being Pinot Noir, the signature wine must be this varietal, and the most well known is the *Hirsch San Andreas Fault Pinot Noir*, named for the famous fault line that runs through the property.

For the tasting, Jasmine escorted me to a wooden picnic table outside, carrying two wine glasses in one hand and a wine case in the other. We sat in the bright sunlight overlooking the green vines, and it was delightful to be surrounded by them, as we tasted the results of the 2010 and 2011 growing seasons.

As she poured the 2010 wine into the glass, the sunlight sparkled off the medium ruby-colored wine with glowing depths of dark pink and crimson. Picking up the glass and holding it over my white writing pad, I could easily see through the lighter colored liquid of a natural Pinot Noir. The thinner skin of the grape meant that it usually produced a paler red wine, unless it was heavily macerated or blended with darker grapes as some winemakers elect to do.

Swirling the wine in the glass to release some of the aromas, I lifted it to my nose and inhaled the black cherry and earthy notes of the 2010 San Andreas Pinot Noir. On the palate the black cherry continued and was joined by a touch of black licorice and a hint of truffle. The wine was muscular with larger tannins than I was expecting, but very satisfying and a wonderful match for roasted pork or duck. Its complex flavor and long length reminded me of a masculine Pommard from Burgundy during a warmer vintage.

"If you remember 2010, it was a strange weather year," Jasmine said. "At first we had a cold spring and summer, and then some very warm weather during harvest."

I nodded, remembering the schizophrenic weather year, and how it was described on the Hirsch website:

"When the pick began on September 21, some of the field test results had the sugar at 21.5 degrees of brix or lower, which indicates final alcohol of 12.5% or below. For the first five days we picked in balmy weather. Then it shot up into the high 80's and 90's. (We pick at night to take advantage of the cool temps to protect the fruit, but one night at 2am during this heat wave the gauge read 82 degrees!) The remaining fruit got ripe all at once and we rushed to pick."

"Are you ready to try the 2011 San Andreas Fault?" asked Jasmine. "Even though 2011 was a cold and difficult year for other regions of Napa and Sonoma, here we had clear skies and consistent temperatures that allowed the grapes to achieve great balance."

She poured the shimmering wine into my glass and I could see it was a lighter ruby red with a touch of purple on the rim announcing its youth. The first notes on the nose were a light raspberry with an intriguing soft floral of rose and violet. On the palate, the raspberry notes broadened out with a hint of spice and plum. I was surprised at how approachable the wine was, but with a clear elegance and crisp acidity that complimented the long finish. Where the 2010 was complex and muscular, the 2011 was elegant and graceful.

"Wow, what clear vintage variation," I said.

Jasmine nodded and smiled.

Later I caught up with winemaker, Ross Cobb, to learn more about the winemaking process for the *Hirsch Vineyards San Andreas Fault Pinot Noir*. Ross has an excellent pedigree in crafting Pinot Noir with previous jobs at both Williams Seylem and Flowers, and a degree in Agriculture and Soil Sciences from UC-Santa Cruz. His voice was filled with earnest enthusiasm as he described the process he used in determining the best time to harvest grapes for winemaking.

"Well as you know," Ross said, "the San Andreas Fault Pinot Noir is composed of multiple blocks. For example, the 2010 is composed of 29 different blocks, and for the 2011 we sourced from 27 blocks."

"What are you trying to achieve?"

"David, Everardo and I spend weeks tasting in the different blocks. Ideally I'm trying to pick when the taste continuum transitions from strawberry to cherry, but not over ripe cherry. There is not a clear brix number we are aiming for because the blocks are dynamic, but in general we usually pick around 22.5 brix."

"What do you mean by dynamic?"

Well," said Ross, "Hirsch Vineyards is not just complex geologically, but the climate is always shifting." His voice was filled with enthusiasm as he spoke, and his words seemed to be propelled forward by an intense energy. "We have an ebb and flow of cold and warm air, and the temperature can shift 20 degrees in the course of an hour. So our picking decisions are often based on the fluctuating fog. If it comes in and covers a block we were planning on picking, we may have to wait a day because the grapes will retract due to the coolness, and we need to let them warm up again."

"Fascinating." This was a concept I hadn't heard before, and another impact of the fog-laden Sonoma Coast that made it such a unique location. "So how does the vineyard and location impact the wine you make?"

Ross grinned. "This is a great location, and I love living on the property and being in the vineyard. No one comes by and drops in on you unexpectedly, so there are no distractions socially. I get to work with a great team, and David and Everardo are the best viticulturists I've ever worked with."

"Tell me more about how you craft the San Andreas Fault pinot."

Ross described the process of careful selection in the field where individual blocks are picked by hand at night. The fruit is then delivered to the winery where a portion is destemmed and inspected a second time on a sorting table, so that any unripe berries or green jacks can be discarded. The other portion remains as whole cluster and is also inspected on a second sorting table to insure the clusters are ripe.

The must is then moved to small open top stainless steel fermenters, ranging in size from .8 ton to 3.5 tons, so that all lots ferment individually. Most lots undergo a cold soak of four to six days, protected by dry ice at an average temperature of 50°F. The cap is generally punched down one time per day during the cold soak to encourage aromatic development.

At the end of the cold soak, the dry ice is removed and fermentation is allowed to start naturally using native yeast. During the height of fermentation temperatures average 85°F. Individual lots are punched down by hand two to three times per day. According to Ross, "Our goal is not to extract color during the punch downs, but to extract flavor. I don't want Pinot Noir to be any darker than ruby."

Once fermentation is complete, which can take anywhere from one to three weeks, the wine is gently pressed using a basket press and then transferred to French oak barrels where it undergoes malolactic fermentation (ML) and is topped as needed.

The wine rests in barrel for 18 months and is not racked or blended until bottling. In early spring the process of blending begins, and Ross and David spend hours tasting the various lots in order to determine the final blend. After blending, the wine is left to marry in tanks for several weeks before it is bottled. The final blend generally includes 30 to 50% new oak, depending on the vintage. The wine is aged in bottle for four to six months before release.

"We make all of our wine quite naturally," explained Ross. "In fact, we generally only use about 40 to 60 parts per million (ppm) of total SO2."

I was impressed, because most commercial wines average around 150 ppm of sulfites. The issue of too many sulfites in wine is one that is of concern to some consumers, though government regulations limit high levels.

"So what is your opinion of these two different vintages of San Andreas Fault pinot?" I asked.

"Very different," he said. "2011 was a perfect year for us, even though it was quite different in the Russian River and other places where they make Pinot Noir. For us it was one of our best vintages with a long even ripening season. But in 2010 we had a lot of uneven ripening, and I was holding my breath, trying to make the best decision on which blocks to pick when." He paused, and when he continued speaking his voice was filled with warmth and wisdom. "But in the end, you just do your best, stay flexible, and listen to the site."

Late Spring & Summer

"Bloom & Verasion"

Stagecoach Vineyard

Bacigalupi Vineyard

Hanzell Vineyard

Chapter Eight

Stagecoach Vineyards
Atlas Peak AVA, Napa County

Stagecoach Vineyards harkens back to the romantic history of Napa Valley in the late 1800's when the swashbuckling bandit, Black Bart, preyed on passengers riding the stagecoach from St. Helena to the town of Monticello. As I drove towards Stagecoach Vineyards on a sunny spring morning, I couldn't help but wonder if Black Bart had traveled the same route over one hundred years earlier.

After driving through the town of Napa, I turned north on the Silverado Trail and watched carefully for the Soda Canyon Road exit on my right, which would take me to Stagecoach Vineyards. Though it was only a couple miles up the road, I still nearly missed the turn and had to swing my car in a large circle to begin the gradual climb up the foothills of the Vaca Mountain Range.

Ten minutes later I passed the ruins of the once grand Napa Soda Springs Resort that opened to guests in 1881. During its heyday the hotel was known for elaborate parties for the rich and famous who came not only to relax in magnificent surroundings with a view over Napa Valley, but to partake of the mineral waters bubbling from the ground that were advertised to heal many ailments. Now all that was left of the ruined resort were two stone entry posts behind a locked gate and fallen rock walls engulfed by trees and shrubs. However, rumors that the old estate was haunted continue to persist to this day.

As the road continued to snake its way up the canyon, I was surprised to see how the vegetation changed from large oak trees, lush green grasses, and orange California poppies at the bottom of the hill to smaller scrub oak, maroon manzanita, and prickly cacti as it rose higher. The soil became tinged with red and large boulders began to appear on the steeper hillsides. The landscape started to remind me of the high mountain deserts of Southern California and Arizona.

Eventually I came to a fork in the road and knew I was supposed to follow the smaller path to the left, which was marked by a bank of mailboxes. Turning down the narrow paved lane that was lined by tall green oak trees on both sides, I realized I had entered a lovely hidden valley. On each side of the tree-lined road vineyards and pastures spread out across the land, and tucked into the hillsides, I caught glimpses of a few houses.

Suddenly my GSP warned that I had arrived at my location. Looking to the left, I realized I had passed a simple metal gate with a call box on a pole. Backing up, I then saw a sign behind the gate announcing Stagecoach Vineyards, and noticed a small red barn like building that was the tasting room and office.

Reaching out to push the button on the call box, I was concerned when there was no response at first, but after a short wait the gate slowly opened. As I drove forward two people appeared on the small front porch of the building, and I recognized the tall, angular figure of Jan Krupp who had the air of a gentleman cowboy about him. He strode forward to shake my hand, and I noticed he was dressed in black jeans and a long-sleeved blue shirt. His abundant silver hair was brushed back from a tanned face, and he wore dark glasses to shield his eyes from the sun.

His grip was warm and firm, and his voice welcoming. "Welcome to Stagecoach Vineyards. I'd like to introduce you to our viticulturist, Gabrielle Shaffer."

Gabrielle stepped forward to greet me, and I was pleased to finally meet a female viticulturist, but her appearance was in sharp contrast to her job. Instead of a tanned and lined face from spending hours in the sun, her complexion was one of white creamy skin with pale blue eyes and short coal black hair. She reminded me of Snow White, and her air of quiet earnestness and delicate movements, only emphasized the impression of an ephemeral creature from another century.

"How wonderful to meet a female viticulturist," I said.

Gabrielle smiled slowly. "Yes, there are not that many of us." She lifted her hand to gently shake mine, and I noticed she carried a large straw sun hat in the other. That and plenty of sunblock must protect her skin, as well as the long-sleeved red and white checked shirt she wore, with a black vest and dark jeans.

"Well, are you ready to tour the vineyard?" Jan asked. When I nodded he motioned to a silver Toyota Land Cruiser parked outside the building and encouraged us to climb in.

Gabrielle insisted I sit in the front seat, while she climbed in the back with a large wine box. "I'm bringing some wine samples that we can taste later in the vineyard," she said.

We exited through the gate, turned left, and continued along a narrow paved road that eventually turned into gravel. Soon we were surrounded by a landscape of coyote bush, chaparral, and the occasional short red manzanita tree. Once again I was reminded of the desert.

"But where is Stagecoach Vineyard?" I asked, wondering where the vines were.

Jan chuckled. "You'll see it in a minute, but this is the road we built to approach it. Many people are surprised we have this type of desert landscape in Napa County."

"Yes," I agreed. In the more than a decade in which I had lived in Napa and Sonoma, I had not encountered this type of terrain.

"There was no road here when we first bought the land," Jan continued. "It took 135 legal documents, signatures from 28 neighbors, and over one year to get permission to make a road, but only two months to build it."

"Amazing," I said.

Jan continued to steer the SUV around curves in the road, a small trail of dust in our wake. Suddenly we rounded a corner, and I could see a huge expanse of green vineyards covering half of a large hillside rising in front of us. I caught my breath in surprise, because the dry desert had been transformed into a green Eden of vines.

"That is Stagecoach Vineyard," Gabrielle's voice floated over the backseat. "We have 600 vineyard acres spread across 1100 total acres of land."

Stagecoach Vineyard

The vineyard was vast, undulating, and shining in the sun. It reminded me of a giant Roman mosaic with the lighter green leaves of the vineyard blocks as the inner pattern and the dark green of the chaparral brush serving as the binding cement. It was massive and mesmerizing.

"Wow." It was all I could think of to say.

Jan laughed. "Yes, it's pretty big, and we had to clear a million tons of rock over the past decade in order to plant it, and we are still expanding."

"But how do you water it? A vineyard in the high desert?"

"Well," said Jan, "as you know vines don't use that much water once their root system is established. Also we have an underground river and aquifer here."

Just then we approached a fork in the road, and in the middle of it sat an historic red stagecoach with giant yellow wheels. It looked just like the stagecoach that Wells Fargo Bank used as its company emblem.

"You have a stagecoach!"

"But of course," said Jan, stopping the SUV in front of the stagecoach so we could see it more clearly. "This is an original one that we purchased a few years ago and had restored. We had to repaint it red, but the yellow wheels and runners are original iron. They painted them that color in the old days. The stagecoach is here to symbolize the history of this region, and the name of our vineyard."

History of Stagecoach Vineyard

In the 1870's German immigrants first settled in the valley and planted crops, including grapes. A road connecting St. Helena to Monticello ran though the valley, stopping at the Soda Spring resort on the way. The road was used frequently to transport passengers in stagecoaches, as well as to deliver supplies, including money to the banks located in the small towns along the way.

Various reports indicate that the stagecoach robber, Black Bart, lived in the hills, and robbed several Wells Fargo stagecoaches along the route. Black Bart was actually Charles Earl Bowles, and he was apparently angry with Wells Fargo because he believed their agents had cheated him out of a mine claim in Montana. His revenge was to rob 28 of their stagecoaches between the years of 1875 to 1883, until he was caught and sentenced to San Quinton prison. He was known as the gentleman robber, because he often left poems and never took money from passengers, only Wells Fargo.

The German immigrants who lived in the valley continued to thrive until Prohibition hit, and they found it difficult to sell their grapes. They also experienced a problem with deer eating the crop, because there was no adequate deer fencing at the time. Therefore, the vineyards were slowly abandoned, and eventually the property was sold to a family that used it as a hunting camp.

In the mid 1950's the Bureau of Reclamation decided to create Lake Berryessa by blocking Putah Creek with a dam. This caused the historic town of Monticello to be flooded and all inhabitants were forced to move. Therefore the old road from Napa Valley to Monticello, where legend says Black Bart used to lay and wait for Wells Fargo stagecoaches to rob, was left to slowly disappear. Today Monticello is a ghost town sleeping at the bottom of Lake Berryessa.

The valley continued to slumber until 1991 when Dr. Jan Krupp saw the property advertised in the *San Francisco Chronicle*. Jan had grown up in New Jersey and inherited a love of gardening from his mother and uncle, which later in life became a passion for wine and viticulture. Upon completion of a medical degree at Stanford University, Jan set up a private practice in Marin County as an internist, and after many successful years working, he began searching for property to plant a vineyard. His dream was to find a site that would produce wines to rival those of the great chateaux in Bordeaux.

After consultation with many experts, he focused his attention on the hills of the Vaca Mountain range where great vineyards were already established, including Dalla Valle, Colgin and Chappellet. When he first found the property, he learned the reason no one else had bought it was because of lack of water and the fact that there were many large boulders that would be difficult to clear. However he noticed a few old vines from the German settlement that appeared to be around 100 years old, and this gave him hope.

Therefore, he hired a geologist to drill a series of water wells to 300 feet, but after the fifth well when there was still no water, he wondered if he should give up the project. Fortunately someone told him about a "water witch" who lived in the valley and could assist him. Using a dowsing rod, she was able to lead them directly to a drilling location. At

400 feet they found an underground river. This led Jan to drill the geologist's sites another 100 feet, resulting in excellent water at all of the locations.

So Jan purchased the 1100-acre property with the support of his brother Bart, and over the next seven years they financed the development of the vineyard. In order to move the large boulders, they had to use bulldozers and in some locations, dynamite. At one point some of the neighbors objected to the project and vandals torched one of the bulldozers. However by taking the time to build positive relationships with the community, Jan and Bart were eventually able to begin planting vines.

Today Stagecoach Vineyard has almost 600 acres of vines with a focus on Cabernet Sauvignon and is the largest vineyard in Napa County. Its location is one-third in the Atlas Peak AVA, one-third in Napa Valley AVA, and one-third in the prestigious Pritchard Hill area. More than 90 wineries purchase grapes from Stagecoach, and there is a long waiting list of customers who desire to purchase more.

Touring Stagecoach Vineyards

"So you named it Stagecoach Vineyards because of the story of Black Bart robbing the stagecoaches that passed through here?" I asked, peering out the window of the SUV to see the historic wagon more clearly.

Jan grinned. "That and the fact that my brother's name is Bart. Besides it sounds better than Krupp Vineyards, don't you think?"

I laughed and had to agree that from a marketing perspective, the name Stagecoach was very attractive as well as easy to pronounce and remember. "So you decided to brand the vineyard?"

"Yes, we knew we wanted to sell the grapes as part of our revenue stream, so we decided to find a compelling brand name." He put the SUV in gear and steered it back onto the gravel road. "But we also produce a small amount of wine under the name of Krupp Brothers, including a Syrah called Black Bart."

As we continued driving I reflected on how rare it was for a vineyard to be branded as part of a marketing and sales strategy. Most vineyards were established as an agriculture crop to be sold to wineries with the appellation as the main driver of price. Then overtime, if the owners were fortunate, the vineyard slowly became recognized as producing high quality fruit and wine. But the Krupp Brothers actually set out to brand their vineyard as part of a business strategy, and succeeded in developing a high quality reputation for Stagecoach within less than a decade.

Historic Stagecoach in Vineyard

"We have wonderful customers that help us promote the vineyard as well," Gabrielle said, leaning forward from the backseat.

"Yes," Jan agreed. "Winemakers like Paul Hobbs and critics such as James Laube have really helped us with good reviews. Also the relationships we have built here in Napa Valley, and word of mouth have assisted us in growing the reputation of Stagecoach. Today we even have customers from Europe who buy our grapes."

"From Europe?"

"Yes, they mainly purchase them for winemaking projects they are doing here in the States," said Jan. He then turned and point out the window. "We are just coming to some of our first blocks. These are primarily Cabernet Sauvignon and some Merlot."

Looking out I could see orderly rows of vines trained on VSP trellising. The bright green leaves moved gently in a soft breeze and I could just make out the white lace of the clusters in full bloom.

"Your vines are in bloom now," I said, excited to see such a beautiful site.

"Yes," said Gabrielle, "and wait until you smell the beautiful flower scent. I love being in the vineyard at this time of the year." Her voice was soft and musical with a clear note of enthusiasm.

We continued driving and I was amazed at the immense size of the vineyard. We passed block after block of vines, with small signs attached to the end posts detailing the varietal, clone and rootstock. The website provided information on the vines, and I knew there were sixteen different varietals planted at Stagecoach, but with Cabernet Sauvignon dominating at more than 55% of the plantings.

"The vineyard is divided into four sections," Jan commented, staring directly ahead through the windshield as he maneuvered the SUV over dips and curves in the road. "Right now we are traveling through the Atlas Peak region, which is the warmest area with a lot of red varietals that produce wines with softer tannins. You can also see that the soil is redder here because it has a higher amount of iron in it."

Just then a California quail ran across the road in front of us with its distinctive black head plume wiggling back and forth before it scuttled into a bush. A small puff of red dusty soil floated out from the bush where the quail had disappeared. Jan slammed on the brakes just in case other members of the quail family decided to cross the road.

"We have lots of wildlife here," said Gabrielle leaning forward. "In fact we have many bird perches in the vineyard as part of our environmental program. We like to encourage hawks and owls because they eat gophers that may hurt the vines. We even have a nesting pair of golden eagles and they had babies last year." Her voice was filled with passion as she spoke and her blue eyes sparkled.

As Jan slowly drove forward again, Gabrielle described more of their environmental practices and answered questions about her background. She had grown up in Minnesota and was half Swedish and half Guatemalan, accounting for her unique coloring. After working back east as a sommelier, she had moved to California to study viticulture through UC-Davis's online program, and then worked at Miner Vineyards before coming to Stagecoach. Her position as viticulturist included a strong focus on vine nutrition and implementing their environmental programs.

"We're now entering the second and third regions of the vineyard," said Jan. Glancing up I realized we were coming into a small valley with vineyards climbing up the hills on both sides. Off in the distance, at the far end of the valley, I knew the land dropped off into steep rugged cliffs overlooking Rector Canyon. At the far bottom lay Napa Valley with the Napa River running through it and San Pablo Bay further south. Both provided a gentle maritime influence and regulated the fog to some extent.

"This hillside on the right is what we call the Heart of Stagecoach," said Jan. "It has more of a south-facing exposure and yellow clay loam mixed with volcanic red soil and large boulders. Because it receives more sun, we plant more of our Cabernet Sauvignon and Syrah there because those varieties benefit from more warmth. Whereas the hillside to the left is what we call the Bordeaux Region, and has a northeast facing slope with deeper clay soils. It was there that the German settlers planted some of their vineyards, because they didn't have to move large rocks."

Looking around I realized the left side of the slope, which Jan called the Bordeaux Region, would actually receive less sunlight during the day and probably had slightly cooler temperatures because of that.

As if reading my thoughts Jan continued. "We call that the Bordeaux region because it has cooler temperatures and is more similar in climate to that region of France. Because of this the grapes from that side usually have a higher acid and more elegance. We plant a lot of our Sauvignon Blanc and Cabernet Franc there, as well as Malbec to help control its vigor."

I nodded, recognizing that these varietals would perform better in cooler areas, and that Malbec, known for its high vigor in terms of producing many leaves and clusters in warmer climates, would benefit from a little less direct sun.

The SUV moved slowly forward with vines surrounding us on all sides. Eventually we entered a small clearing with a cluster of buildings. Grape picking bins were stacked outside.

Jan Krupp

Gabrielle Shaffer

"That's my office there in that brown trailer," Gabrielle said, pointing at a long building to the left of the road. "I have my computer in there and can track all of the information coming in for each block."

"Yes, we keep Gabrielle plenty busy here," said Jan with a smile, as we drove past her office. "One of the projects we're working on is a new planting of Cabernet Franc that you'll see in just a minute."

True to his word, a few minutes later we came to a vast field of bright blue milk cartons. They stood out in sharp contrast to the red soil, but I knew they were protecting newly planted vines from fluctuating temperatures, wind, and predators such as rabbits that may eat the tender young leaves

"Our employees plant the vines, and our vineyard management firm provides the protective cartons," explained Jan. "That way they can recycle them for other projects."

At the end of the milk carton block, Jan turned the wheel and we headed up a steep hill into the Bordeaux Region. I hung onto the sides of my seat as the SUV surged upward at a sharp 30 degree angle, passing row after row of carefully tended vines. He stopped at the top of the hillside near a block of vines that were in full bloom.

"Let's get out here for a minute," said Jan.

As soon as I opened the door of the vehicle the scent of sweet perfume filled the air, and I knew I was immersed in the magical few days each year when a vineyard is in full bloom. Sniffing the air, I tried to name what I smelled, but the aroma was very delicate and powdery. Perhaps closer to the scent of a white lily just before it opens or honey-suckle at the end of its cycle.

"Come and see the flowers."

I looked around to see Gabrielle in the midst of the vineyard with the large green leaves of a Cabernet Franc vine covering her upper body as she bent her nose into a fragrant cluster. I rushed over to her, while Jan stood next to the truck and grinned at us.

Gabrielle gently cupped a cluster of the Cabernet Franc blooms in her hand, and as I gazed down at them I was enchanted to see the diminutive white petals that always reminded me of tiny stars in a miniature galaxy. It was fascinating to realize that each flower would eventually become a grape.

"I especially love the scent of Cabernet Franc flowers," said Gabrielle with a gentle smile on her face. "They smell like lilies to me, and see how long and delicate the cluster is."

She was right. The length of the cluster did appear longer than others I had seen, especially the smaller Pinot Noir clusters in my home vineyard. "Absolutely exquisite," I said.

As we stood in the block, I noticed that all of the vines were on VSP with double cordon and spur pruning, which is traditional with many Bordeaux varietals, such as Cabernet Franc and Cabernet Sauvignon. The moderate width of the trunk suggested the vines were about ten years of age, and appeared healthy and verdant.

"Are you primarily using VSP with spur pruning for all of Stagecoach?" I asked.

She nodded. "For the most part, but it depends on the client."

Jan walked forward to join us. "We have a few blocks that are on uni-lateral cordon," he added, "and some older ones that are quad, or four cordons per vine."

"But we're starting to switch to cane pruning on some blocks," said Gabrielle, "because it helps reduce *Eutypa*." I knew she was referring to the disease that causes sections of the cordon to die due to a fungal infection usually caused by large pruning wounds. With cane pruning, the cuts were smaller and there was less chance for infection. "We have a block near-by where we are experimenting with this," she continued.

"Let's go check it out," said Jan.

We climbed back in the truck and a few minutes later stopped near another block of much older vines. Jumping out of the vehicle I could see that sections of the cordons had been chopped off due to *Eutypa*.

Gabrielle walked up to one of the vines and ran her fingers along the amputated limb. "You see, we had to cut the cordon here, but we are grooming these two shoots to be canes." Her fingers touched two new shoots covered with leaves that were growing from the trunk of the vine near the training wire. "The rootstock of this vine is still alive, but the cordons are dying. It still produces good quality, but less fruit, so we are slowly switching to cane pruning here and elsewhere because it has less Eutypa problems."

"We started with unilateral cordon on VSP," added Jan, "but now we realize there are less *Eutypa* issues with cane pruning."

"Interesting," I said, "and Stagecoach is a relatively young vineyard. What year did you plant your oldest block?"

"1996," said Jan.

"So what is your opinion of the quality of fruit coming off younger vines?" I asked.

Bloom on Cabernet Franc Vine

"We've found that some of the best wine comes at six to eight years," responded Jan, "but we have some people who want to pick at three years, because we are planting dormant bench grafts so the fruit is ready earlier. Also since the customers have to pay for a new vineyard block, many decide they want to harvest in three years so they can make a little wine. However, if the vine is less than three years old, we prune to one cluster per shoot to insure the vine has the energy to ripen it."

"Gabrielle, what is your opinion?"

She sighed softly before responding. "There are many different opinions on vine age. If you ask a group of winemakers, they will all say different things."

Jan laughed. "Isn't that the truth, but one of the things we have learned over the years is about the difference in the soil. For example here in the Bordeaux region where the Germans were able to plant vines, the soil is deeper because it is made of volcanic tufa, like this."

He bent down and picked up a thin rock that was about two inches in diameter. It was pale yellow with red and orange streaks running through it. He held the rock between his thumb and forefinger in both hands, and then quickly snapped it in half. "You see," he said, holding the two halves of the rock out to me, "it breaks easily. This type of rock is softer, like sandstone, and grapevines can push their roots through this, but in other parts of the vineyard the soil is very shallow over hard bedrock, so we need to dynamite through that."

"Can we visit a section where you've had to use the dynamite?"

"Of course! That's where we're headed next."

Back in the SUV Gabrielle handed around bottles of water and we all took a long drink. Jan steered down an even steeper hill this time with some large ruts and boulders, and I watched as he shifted into four-wheel drive. Once again, I clung to the edge of my seat as we plunged downward at a 40-degree angle, and was glad the seat belts appeared to be working well.

"So we're four-wheel driving," I said.

Jan had removed his sunglasses and when he turned his head quickly to look at me, attractive laugh lines fanned out from his dark eyes and his teeth flashed white as he grinned broadly. "Yes, that's half the fun of working here, and you know we offer four-wheel drive tours through the vineyard for tourists."

"Another great way to promote the vineyard brand," I said.

Jan focused on steering the vehicle around a hairpin turn in the road before responding. "You bet, and people seem to enjoy coming out here."

"We've even had a couple ask if they could have their wedding here," Gabrielle commented from the backseat.

"I can understand that." With the green vineyards spreading out around us, the mountains rising up behind, and the Napa Valley below, I had to agree it would be a lovely place to exchange vows.

We were approaching a field of brown soil with a pile of rocks and boulders forming a mound in the middle. Sprinkled throughout the dirt were orange cones similar to those used in highway construction projects. Behind the field was an even larger mass of boulders.

"So this is a field that was recently dynamited?" I asked.

Jan stopped the SUV at the edge of the dirt field. "Yes, we are now in the last section of the vineyard which we call the Pritchard Hill Region. As you can see there are many large boulders here, and below the topsoil it is solid rock. If we don't break up the rock, the grapevines cannot push their roots through to grow."

Looking around I had to agree that the terrain looked very rugged, with red and cream-colored rock outcroppings and desert sage with chaparral brush. Again, I had the impression that we were in the high desert, and it was hard to believe the lush green beauty of the Napa Valley was just below us several miles.

"So what are those orange cones?"

"The cones signify places that will be dynamited next," he replied. "Do you see that pile of rocks in the middle of the field?" He pointed through the windshield. "That is what we've dynamited so far."

As he explained the process, I was amazed at how complicated and labor intensive it was. First they drill two to three hundred holes about ten to twenty feet deep into the solid rock. Next they put dynamite in each eight-inch diameter hole to blow it up. The soil and rocks shoot about eight feet in the air and come down in a big cloud of dust. Then they use both an excavator and bulldozer to move the rocks to the mound in the middle of the field, and rip the soil to about three and a half feet. Eventually the small mound of rocks will be transferred to the larger pile of boulders on the hill behind. These are then sold to landscape designers and other people who want to buy them, while some of the boulders are ground into gravel to use on the roads.

Preparing New Vineyard for Dynamite

"That is an amazing process," I said. "How much of Stagecoach Vineyard has been dynamited to date?"

"Only about 25%, or 150 acres," said Jan. "The soil in Stagecoach is diverse. As you saw, the soil in the Bordeaux section is deep with decomposed rock that breaks easily, so no dynamite is needed there. In other parts of the vineyard we have huge boulders the size of cars that we move with bulldozers, but here in the Pritchard Hill Region we have more solid rock underneath that requires dynamite."

"Well," I said, "playing the devil's advocate, I'm sure there are some people who would ask why you are even planting grapes here if you have to dynamite the land in order for them to grow. What not just plant elsewhere?"

"Good question," said Jan, "and the answer is if you want amazing high quality wine with structure and great aging potential, then this works. The end result here is well-draining soil with unique properties and lack of nutrients, that when combined with this special climate, creates balance, power, and complexity in the final wine."

As Jan spoke, his voice was filled with emotion, his eyes flashed with light, and the distinctive black eyebrows that contrasted so well with his silver hair were raised in alert attention. As I listened to him, I realized that this was a man with a clear vision and the perseverance to overcome many obstacles, including solid rock, dissenting neighbors, and a lack of water when he first set out to establish the vineyard.

"So how do the neighbors feel about the dynamite?"

"They've actually been quite understanding except for once when we dynamited on a Sunday. They didn't appreciate it so we stopped."

"I can understand how they might object," I said. "So who actually sets the dynamite?"

"One of our drivers had a license for dynamite," Jan responded, "so we started doing that to loosen the rocks, but I had to hire someone else because he was arrested for pulling a gun on the sheriff."

"Really!"

Jan grinned, and his eye sparkled. "Yes, that really happened, so I had to hire another licensed professional to handle the dynamite."

"Are you ever involved?"

"No," Jan said and then grinned even more broadly, "but I like to drive the bulldozers."

Vineyard Specifics for Stagecoach

Departing from the dynamite field, we headed toward the lookout on the extreme edge of the Heart of Stagecoach region. As we traversed the gravel roads, Gabrielle and Jan provided more information about the specifics of the vineyard.

The total planted 600 *acres* of vineyards are comprised of 16 *varietals*, of which the greatest percentage is devoted to the five red Bordeaux varietals: Cabernet Sauvignon, Cabernet Franc, Merlot, Petit Verdot, and Malbec. At the same time, they also have a mix of other grape types including the Rhone varietals: Syrah, Grenache, Marsanne, Roussanne, and Viognier. The remainder comprise an international mix of Tempranillo, Sangiovese, Zinfandel, Petite Sirah, Sauvignon Blanc, and Chardonnay.

Though Stagecoach is most famous for its powerful Cabernet Sauvignons, they also have a reputation for producing excellent Petit Verdot. "Michel Roland," said Jan, "the famous French winemaking consultant, told one of our customers that our Petite Verdot was the best he had ever tasted."

The *soil* is primarily Hambright complex with small amounts of Guenoc-Rock outcrop. As Jan mentioned it is different in certain sections of the vineyard, ranging from the solid rock of granite and basalt, to clay and clay loam, and to the deep soils of decomposed volcanic lava in the Bordeaux section. In fact this latter type of soil, which is residual bedrock created by water eroding large boulders, is so unique that it was described in detail in the book, *The Winemaker's Dance: Exploring Terroir in the Napa Valley:*

> *"We found an extraordinary example of this phenomenon at Stagecoach Vineyards,...When Jan Krupp developed Stagecoach, he found what he believed were soils up to sixteen feet thick, with boulders as large as twelve to fifteen feet in diameter....(we) identified the boulders as core stones, which had been formed in place by the weathering of an originally solid layer of lava (p. 80)."*

The *elevation* of Stagecoach Vineyard ranges from 1050 to 1800 feet above sea level. According to Jan the higher elevation creates a unique location in the Napa Valley because it is cooler, but has more sun and less fog. He compares it to the high desert vineyards of Washington State, with more hours of sunlight.

"Yes," Gabrielle agreed. "I live in Napa Valley and I enjoy driving up the hill to Stagecoach each morning. When I leave Napa, it is often foggy, but when I arrive here and emerge into the valley, I break into the sunlight."

Vineyard Specifics for Stagecoach

Total Vineyard Acres	600
Varietals	Cabernet Sauvignon, Cabernet Franc, Merlot, Petit Verdot, Malbec, Syrah, Petite Sirah, Grenache, Tempranillo, Sangiovese, Zinfandel, Sauvignon Blanc, Chardonnay, Marsanne, Roussanne, Viognier
Soil	Hambright complex with small amount of Guenoc-Rock outcrop
Elevation	1050 to 1800 feet above sea level
Average Temperature	Summer temperatures average 85° F for a high and 55°F for a low at night. Rainfall is usually around 34 inches
Rootstocks	Half of vineyard on 110R; rest on 101-14, 1103, St. George, 140, 3309
Clones	62 different clones. Primary clones for cab: 2, 4, 7, 8, 15, 29, 31,Weimer, 337, 412, and 685
Sun Exposure	South to Southwest
Spacing	Primarily 6 x 4, but some 7.4 x 6 on older vines
Trellis Systems	VSP with both spur and cane pruning

Average *temperature* in the summer is 85°F during the day and 55°F at night. Interestingly due to their higher elevation, they often have light snow fall in the winter. Rainfall is usually around 34 inches.

Stagecoach Vineyard is planted on a variety of *rootstocks*, all carefully selected to match the different soil types. "Over half of the vineyard is on 110R," reported Jan, "because it is drought tolerant. We are a desert and want to preserve water." However in other cases they have planted 1103, St. George and 140 for rockier soils, whereas 3309 and 101-14 are used in deeper clay based soils. "These types of rootstock perform better in these different types of soil," said Jan.

The wide variety of *clones* at Stagecoach is quite amazing. Currently they have 62 different clones for the 16 varietals. "Our customers want a selection of clones from which to choose," explained Jan. "Some clones produce unique flavors and characteristics that winemakers find desirable." Clones they are using for Cabernet Sauvignon include: 2, 4, 7, 8, 15, 29, 31, Weimer, 337, 412, 685, and several others.

Vine orientation is mainly north by south, because it provides the best sun exposure, and is easier for the vineyard tractors to travel up and down the hill. However Jan admits that some winemakers prefer to have the rows running east to west lengthways along the hills.

"Though this is more difficult for the tractor drivers," explained Jan, "some winemakers believe it is better for their fruit in terms of canopy management because it provides more shade to the clusters. However we find we often have to pull leaves on the shady side, which causes vineyard labor costs to increase." He paused and then smiled, "But since we are planting these blocks to please the customer and they pay for the installation, we gladly make these types of adjustments."

"Interestingly," said Gabrielle, "the old timers I've talked to around here told me that it doesn't really matter which direction you plant the vines, but instead they recommend matching the rows to the contours of the hills."

"Yes, I've heard that same philosophy as well," I said, thinking of some of the older vineyards I had visited. "I wonder which perspective is correct?"

Spacing at Stagecoach is primarily 6 x 4, but some of the original blocks were staked out at 7.4 x 6 with 990 vines per acres, according to the philosophy of the time. There is only one type of *trellis system*, VSP, which Jan believes to be the most practical and useful for both the vines and the vineyard workers. However, they are utilizing both spur and cane pruning with the VSP, and experimenting with unilateral, bilateral and quadrilateral spur pruning.

Farming Practices at Stagecoach

Stagecoach Vineyard is *certified* by Fish Friendly Farming and practices sustainable farming methods. As part of this focus, they *fertilize* with fish fertilizer as well as local restaurant food waste for compost. This is based on data from petiole analyses that are conducted at bloom and verasion. Occasionally they will apply non-organic fertilizer through the drip irrigation system, and may use foliar sprays.

"We also use cover crop," commented Gabrielle, "both as a partial fertilizer and for erosion control. With the soil only a foot deep in some locations, it is very important to use cover crop to help prevent erosion."

"Indeed," said Jan, "our farming permit actually requires that we agree to use a permanent cover crop because of potential erosion issues. We don't want the top soil flowing down the canyon in a bad rain storm."

Gabrielle nodded in agreement. "And another benefit of the cover crop is it helps reduce vigor, because the vines must compete with the cover crop. This promotes higher quality in the grapes."

Part of their commitment to supporting the environment is a large-scale project they undertook to replant some of the original vegetation in the area. This includes the rare *holly-leaf ceonothus* that attracts honeybees. They have also released beneficial insects to promote a healthy natural environment.

For *weed control*, the vineyard workers mow within the rows and will hand shovel tough weeds. They also apply Round-up if necessary. *Canopy management* is almost 100% by hand, but they will perform some mechanical hedging.

Farming Practices at Stagecoach

Certifications	Fish Friendly Farming Certified, using sustainable farming practices
Fertilization	Cover crop, fish fertilizer and organic compost, & non-organic fertilizers if needed
Weed Control	Mowing and Round-up
Canopy Management	Almost all by hand, but will perform some mechanical hedging; depends on client requests
Disease Control	Sulfur and other organic and non-organic fungicides to control mildew. Monitor for oak root fungus and red leaf virus.
Pest Control	No major issues, using raptor perches, owl boxes, bird feeders and predatory insects.
Irrigation	Drip irrigation. Have 15 wells
Technology	Weather stations, wind turbines, neutron probes
Harvest Measurements	Winemaker clients visit vineyard and determine by taste and lab measurements

"Most of the canopy management work," explained Gabrielle, "is determined by the specific farming plan negotiated by the client."

The annual farming cycle includes pre-pruning in January with completion of pruning in February and March. In April and May they are focused on suckering and shoot thinning, and will de-leaf on shady sides of the vine and pull lateral shoots when the canopy is full in June and July.

"We may also do tunneling," announced Jan.

"Tunneling?"

"Yes, it is deleafing in the center of the vines to open up the canopy so that the clusters can get more light."

In July they will conduct a green harvest of excess clusters if necessary, and/or remove wings and break-up clumps. This type of fine-tuning is continued through verasion. Harvest starts in September, and they generally finish in late October.

"We start harvest a little later up here than the rest of Napa Valley," said Gabrielle. "That is because we are at a higher altitude so it is cooler, even though we have more daylight hours of sun throughout the growing season."

For *disease control* they monitor the vineyard and use organic and non-organic products for prevention. "For example," said Jan, "though powdery mildew is rarely an issue for us, we will apply sulfur and organic products every 14 days and stop at verasion."

"We also utilizer mapping for disease and assess the whole block to determine replants," reported Gabrielle.

"What might cause a block replant?" I asked.

"Red leaf virus can be an issue here, just like the rest of Napa Valley," said Jan. "So if we get too much in a block, we will pull out the vines and replant."

"Sometimes oak root fungus can be an issue," said Gabrielle softly. "It happens if the location had a lot of scrub oak before it was planted."

"What do you do about it?"

"We try to make sure the land is well-cleared of any oak roots before we plant," said Jan. "We're also experimenting with different rootstocks that may be more resistant to it, but in some cases we may have to replant a section of the vineyard."

There are no major issues with *pest control*, except for dust mites on occasion. To combat mites, they distribute gravel on the roads to minimize dust and also use Dust-Off, which is a salt-based product.

"Do you have any problems with birds eating the grapes?" I asked.

"Birds?" repeated Gabrielle, and then she smiled and her blue eyes softened. "No, we invite birds to the vineyard by putting up bird feeders. We find that if we feed them, they don't eat the grapes. In fact with our raptor perches, owl boxes, and feeders, this is a great place for bird watching."

"Yes," agreed Jan. "We've had people on our four-wheel drive tours stop to watch the golden eagles."

"What about other pests such as deer, raccoon, turkeys, or gophers?"

"We have 5.5 miles of deer fencing," replied Gabrielle, "plus three dogs that scare away most other predators. For gophers, we have the hawks and owls that patrol the vineyard, and a few snakes, such as gopher and king snakes, that will eat them."

"Any rattlers?"

"Occasionally," Jan frowned. "We watch for them, but last year two women found a rattler in a vine when they were suckering. They moved quickly out of harm's way, but we do have to be cautious."

"However," said Gabrielle softly, "even though we try to protect employees and the vines from pests, we still want to promote wildlife in the vineyard. We've actually had sightings of mountain lion, bobcats, coyotes and even a bear."

For *irrigation*, Stagecoach is on an extensive drip system and has 15 wells to provide water from their underground river and aquifer. "We don't really use that much water," said Jan. "We only irrigate as necessary to maintain the desired level of stress. This is very dependent upon the specific blocks and varietals. In some areas that may be three times per week and in other areas three times per year. We also monitor the depth of the wells and they have not changed even though we have 600 acres of vines."

There are also different types of *technology* in the vineyard to monitor the health of the vines. This includes 19 weather stations as well as gas-powered fans that automatically turn on if there is a frost alert. In addition to petiole analysis, they also conduct sap flow measurements, use leaf porometers, and apply pressure bombs to measure moisture. Once a year they will hire a company to do flyovers and create NDVI maps (Normalized Difference Vegetation Index). Gabrielle carefully tracks and analyzes all of the data with her computer system.

Harvest measurements are determined by the individual winemakers and their staff who visit the vineyard to taste and analyze the grapes.

Economics of the Vineyard

The *average yield* per acre for Stagecoach Vineyards varies by the individual customer contract and, in general, is quite low. Production also varies by year, but the average is around 2000 tons.

Annual *farming costs* also vary by customer contract, with the average being $11,000 per acre. However installing a new vineyard can cost as much as $250,000 per acre if the land must be dynamited and cleared of large boulders.

Economic Viability of Vineyard

Average Yield	Varies by customer contract
Total Average Tons Per Year	2000
Farming Costs Per Acre	$11,000 per acre
Revenues	Varies by contract; range is $4000 to $15,000 per ton
Economic Health	Very good

Vineyard contracts are established with each customer to document the negotiated farming plan and grape price the customer will pay. According to Jan, "We generally do evergreen contracts of three to five years, but in some cases will do eight to ten years with an option to renegotiate. Contracts may be established to sell by the ton, by the acre, or by the bottle."

In order to support their more than ninety customers, Stagecoach has 100 fulltime employees and an additional 70 they hire through a vineyard management firm. Esteban Llamas, their vineyard manager, and foreman Jose, oversee the work crews. Wages average $13 per hour, but escalate during harvest when the employees work in teams and are paid by the ton, plus potential for bonuses.

"Harvest is our busiest time of the year," reported Gabrielle. "We have to hire extra employees. Last year we had 250 employees per day during harvest. One day we picked 250 tons for 49 different wineries."

"Yes," Jan agreed, "harvest is a crazy time. Also, we don't do night picks because we need to see the clusters to insure quality. Therefore we start at first light and usually finish around noon, but then we also deliver the grapes to the wineries, so it is a lot of work."

In terms of *revenues*, Stagecoach achieves $6500 to $15,000 per ton for Cabernet Sauvignon, depending on the farming requirements for the customer. Other varietals range from $4000 to $10,000 per ton. Some customers have negotiated contracts by the acre, ranging from $30,000 to $40,000 per acre.

According to Jan the *economic health* of the vineyard is very good. "We have a waitlist of customers who want to purchase our grapes," said Jan. "Even though we financed the vineyard, and are still on a payment schedule, we are very confident in our economic health."

Another small revenue stream results from their four-wheel drive vineyard tours. Customers pay $35 for a one and a half hour tour that includes a winetasting in the vineyard.

"We're almost to the site where we do our winetastings," said Jan with a smile, as he steered the SUV around another sharp turn. "We're back in the Heart of Stagecoach section now, but driving to a lookout over Rector Canyon where we have some tables set-up."

Soul of the Vineyard – "Excellence and Extremes"

Just then we crested the top of the steep hill and I caught my breath at the grandeur of the view. The vineyard blocks were behind us, and we seemed to be perched on the edge of steep canyon that plunged hundreds of feet down in a cascade of rocky terraces and shrubs. In the far distance I could see the shimmer of San Pablo Bay.

View from Heart of Stagecoach Lookout

"Magnificent," I breathed in appreciation, quickly unbuckling my seat belt to climb out of the vehicle. Walking towards the edge of the cliff, I was relieved to see it was bordered by huge slabs of rocks and boulders in shades of red and yellow. They not only provided a degree of safety, but looked like a work of art piled artistically along the rim of the canyon.

"That's Oakville below us there," said Gabrielle as she came up to stand next to me.

She pointed at the valley below and I could just make out the small buildings that comprised the hamlet of Oakville. Surrounding it the rest of Napa Valley spread out in a patchwork quilt of green vineyards and golden fields, with the blue green of the Mayacamas Mountain range rising up on the far side.

"On a clear day you can even see San Francisco," said Jan joining us. He carried the wine case from the back seat of the SUV. "So what do you think of our tasting room?"

"Wonderful! I actually think I can see some of the tall buildings in the city," I said, shielding my eyes to gaze into the distance. The faint outline of tall skyscrapers seemed to rise against the pale blue of the sky.

After looking at the view for several more minutes, Gabrielle beckoned us to a picnic table shaded under a large umbrella. It was situated with a view of the canyon, but also near a vineyard block labeled *K8A Heart of Stagecoach Cabernet Sauvignon*. The vines were young and slender with healthy green leaves and fragrant grape clusters in full bloom.

As we took our seats around the table, I asked, "Before we begin the winetasting, can we talk about the personality of the vineyard? What one word would you use to describe it?"

"Excellence," said Jan immediately. "We aspire to grow the best grapes in the world, and contract to sell our grapes to the most talented winemakers in the world. We are a *Grand Cru* vineyard of Napa Valley."

"For me it's a little different," said Gabrielle slowly. "The one word that comes to mind is 'extreme.' It is an extreme site with extreme weather such as snow and heat. It has extreme beauty with the rocks and valley." She paused and looked at Jan with a sly smile on her face. "And he was extremely crazy enough to build up here, and now we've attracted customers with extreme personalities who have dreams to craft excellent wines.

"Extreme excellence," I said. "That's a unique combination. So what is the best and worst part of working here?"

Jan smiled and his eyes twinkled. "The best part is working outside and seeing the growth of the vineyard. I'm here four to six days a week and really enjoy the continuous improvement of making this vineyard great. One of my favorite times is harvest when it is all hands on." He paused before continuing. "On the downside, I'd have to say a late harvest with rain. We've had that a couple times and it was awful. Esteban and I were here constantly trying to figure out how to save the crop."

"What about you Gabrielle?"

Gabrielle had a dreamy look on her face as she answered. "The best part of working in the vineyard for me is growing things. I love plants and farming, and grew up on a farm like Jan." She paused and then let out a soft musical laugh. "I get to be a girl about two days a year when I put on a dress and nice shoes to taste wine with customers, but the rest of the time I am outdoors wearing jeans, hats and sunblock. I love it!" She smiled and her eyes glowed a soft baby blue.

"Seems as if you've found a beautiful place to work," I commented.

"Yes," she beamed. "And I always take time out each day to look around and appreciate the beauty."

"What's challenging for you working here?"

Gabrielle shrugged her shoulders and grimaced slightly. "The balance of trying to make everyone happy. I am a Libra and crave balance. I want our customers to be happy, but I can't always accommodate all their needs. Sometimes I feel like the camp counselor."

"So what have you learned working in the vineyard?"

Gabrielle leaned back and her eyes roamed over the block of Cabernet Sauvignon behind us. "I am a student of this place," she said very softly. "I'm learning about the soils and weather patterns, the rocks and different needs of the plants. It is very fulfilling to be constantly learning."

"Yes," I agreed. "Then life never gets boring. What about you Jan?"

"I've learned to respect the land and not fight with nature," he said in a serious tone. "I've learned how to water and fertilize to match the needs of the vines. We've made many changes over the years such as switching to more cane pruning, tighter spacing, and grafting over some varietals. We've come to realize there were places in the vineyard where Cab wasn't as good, so we changed that. Cab wants more sun. We've discovered Malbec and Merlot are more forgiving of the deeper soils and need less heat to ripen." He paused and looked around the landscape before continuing. "And we've added many more customers by learning how to give them the best grapes we can grow."

"So would you like to taste some of the wines that were created by this vineyard?" Gabrielle asked, pulling out five bottles of wine and several glasses from the case.

As she began the delicate process of prying the cork out of the bottles, I got up and wandered over to the edge of the Cabernet block. Reaching out to touch a leaf, I enjoyed feeling its slightly bumpy texture under my fingers. But I could also feel something else – a throbbing energy that seemed to manifest itself across the whole vineyard.

At the risk of too much alliteration, I thought, I would like to add another "e" to Jan and Gabrielle's equation of "excellence and extreme." The 'e' would stand for 'energy,' because I felt it pulsating through this site. Perhaps it was because there were many young vines here striving to grow, or perhaps it was because part of the very foundation was being rocked by dynamite. Then again it could be the hundreds of people working in this vineyard and the more than ninety customers that visited each year, bringing their ideas, thoughts and skill sets to shape the vines. The abundant wildlife could also contribute to the sense of exuberant energy.

As I turned to go back to the table, another thought struck me. Perhaps it was all of these factors plus the history of the place. The German immigrants who came to settle the valley in the 1800's, Black Bart roaming the hills, and the old stagecoaches that used to rumble over the now faded road to the drowned town of Monticello. All of that energy could be combined here in this hidden valley high up in the Vaca Mountains. The very shape of the hills, forming a gentle bowl could be a container for this special vitality, capturing and generously spreading the energy, before the land dropped off and plunged into the canyon below.

Signature Wine: Krupp Brothers Cabernet Sauvignon

There are many famous wineries producing excellent wines from Stagecoach Vineyards, with several of them receiving high ratings ranging from 92 to 97 in *Wine Spectator* and *Wine Advocate*. I was fortunate because Jan and Gabrielle had brought along a selection of different brands for our tasting.

"The wines are ready to taste now."

Gabrielle's voice floated softly across the vineyard, and I turned to see her seated at the picnic table with the green cliffs of the canyon etched sharply behind. The sunlight sparkled on five wine bottles and glasses lined up in a row on the table. Jan sat next to her with a huge smile on his face. "I think you'll enjoy some of these," he said.

"I'm sure I will." Taking my place at the table opposite of them, I realized they didn't have any tasting glasses. "Aren't you tasting?" I asked.

"We've both tasted these many times," replied Jan, "so we'd prefer to get your opinion. We decided to bring five wines from the 2007 vintage because it was an especially good year, and though warmer than other vintages, I believe it highlights the power of Stagecoach fruit, regardless of the different winemaking styles."

"What a good idea." Glancing at the labels I was pleased to see a Krupp Brothers Merlot and Cabernet Sauvignon, Conn Creek Cabernet Sauvignon, Paul Hobbs Cabernet Sauvignon, and a JC Cellars Syrah.

"In what order would you like to taste?" asked Gabrielle.

"I think you have them lined up perfectly, because I'd like to start with the lighter Merlot, then taste the four Cabernet Sauvignons, and end with the more tannic Syrah."

"Good method," Jan smiled.

As Gabrielle poured the wines, I asked Jan a question. "So can any winery that is lucky enough to purchase your grapes put Stagecoach Vineyard on the label?"

Jan quickly shook his head in the negative and his black eyebrows came together in a frown. "No. Our primary focus here is excellence, so customers can only use the Stagecoach name if their wine passes a taste test and is priced correctly. They must send samples and pricing information, and we make a decision to approve or not."

"Good quality control." I noticed that Gabrielle had filled each glass with about two ounces of the glowing ruby liquid, making sure to line them up in front the correct bottle in the tasting order we had discussed. As the sun bounced off the red wine, I could see that the Merlot was a softer red-garnet shade compared to the darker crimson hues of the four Cabernet Sauvignons, while the Syrah stood out in its opaque inky blackness.

Taking my time, I tasted through the wines, carefully spitting them on the ground. The Merlot was very classic in that it had softer tannins and notes of plum, fruitcake and a touch of herbs. The Cabernet Sauvignons exhibited a range of cassis, cedar, and herbs with all of them expressing the massive tannin structure for which Stagecoach Vineyards is famous, but with a pleasing velvety texture overlaying its muscularity. Though the acid and alcohol levels differed, based on the winemaker's preference, all wines exhibited both balance and power, and were very well made.

The Syrah was a special delight, with its pitch-black depths giving off aromas of tar, wild sage and blackberry that followed through on the palate. It was extremely concentrated, with huge hairy tannins, and a long lingering finish. All I could think to say was "yum."

"So which is your favorite?" Jan asked.

"They are all magnificent, with the Cabernets and Syrah definitely highlighting the tannic power of Stagecoach."

"But which is your favorite?" Jan persisted.

"Well, since I'm talking with one of the Krupp Brothers, let me say how impressed I am with your Cabernet Sauvignon." This was true because the 2007 Krupp Cabernet Sauvignon was especially distinctive showing both the powerful tannins of the vineyard but with an added elegance from the perfectly balanced acid, dark fruit and well-integrated spicy oak.

"That was made by our previous winemaker, Nigel Kinsmen," said Jan. "I'll be sure to introduce the two of you."

As we made our way back down the mountain, winding through the many blocks of vineyards, I felt grateful to visit this spectacular place. It was truly one of the magnificent vineyards of Napa Valley, and the hospitality of Jan and Gabrielle was much appreciated.

True to his word, Jan connected me with Nigel Kinsman at a later date, and we were able to discuss the winemaking methods for the *2007 Krupp Brothers Cabernet Sauvignon*. Nigel has an impressive background, having received his enology degree from University of Adelaide and then making wine at several estates in the Margaret River region of Australia before moving to Italy to make wine with Michel Rolland. Arriving in California, he became winemaker at Jan and Bart's Krupp Brothers winery for several years before moving onto work at Araujo Estate.

"So what do you think of Stagecoach Vineyards?"

"Stagecoach is a truly amazing property," answered Nigel, his voice filled with enthusiasm and his Australian accent clearly evident. "It is rugged, wild and one of the last areas of Napa Valley to be developed. Jan had incredible vision to go to the lengths he did in order to develop it."

"Yes," I agreed. "It is truly amazing. "What do you think about the grapes produced by Stagecoach?"

"The wine is coming out of the rocky ground," Nigel responded, with passion ringing in his voice. "I think this contributes to how large and tannic they can be at times. Stagecoach produces wines that are incredibly interesting and have huge future potential."

"What was the 2007 vintage like?"

Nigel grinned, and his small black goatee moved up and down with his changing facial expression. "I remember it being a very long harvest. The Cab Franc and Merlot were early, but the Cabernet Sauvignon was very late. Those blocks in Stagecoach have very small berries that need time for the tannins to resolve, so we had to wait until it was perfectly ripe. We didn't finish picking until late October."

Nigel explained how they transported the grapes to a custom crush facility at Laird Winery where Jan had purchased and installed his own stainless steel tanks. The grapes went through a cold soak for 2 to 5 days, and then they let the must warm up so that indigenous yeast could start the fermentation.

"I like to keep the ferment at cooler temperatures so not to extract too many tannins," explained Nigel. "Therefore I started it at around 77 degrees, but didn't take it higher than 84 degrees." He paused, and then continued with a laugh. "You are always battling those magnificent tannins with Stagecoach."

With some tanks he performed a short extended maceration, but not if the tannins were too high. "It depended on the tank," said Nigel. "The total maceration times ranged from 24 to 30 days."

The wine was pressed using a pneumatic press, but only five percent of the pressed juice went into the final blend. The wine was aged in 86% new French oak barrels for 22 months, and then blended to create the final cuvee of 92% Cabernet Sauvignon, 5% Cabernet Franc, 2% Petit Verdot, and 1% Merlot.

As we talked I had palate memories of the beautifully structured wine with its aromas of cassis, violets and earth, and powerful velvety tannins, with the long and graceful finish of blackberry and cedar. It truly reflected its "rugged and wild" terroir as Nigel so aptly described, but also expressed an elegance and depth of concentration that the hand of the winemaker contributed to the final creation.

Chapter Nine

Bacigalupi Vineyard

Russian River Valley AVA, Sonoma County

Westside Road is often called the "most scenic wine trail in Sonoma County," and on a bright May morning with the sunlight percolating through shady trees that line the road in some spots and casting rippling patterns of light, I had to wholeheartedly agree. The two-lane road winds its way south along the Russian River starting just to the west of the charming town of Healdsburg and ending at the intersection of Guerneville Road. My destination was Bacigalupi Vineyard, which lies approximately halfway along Westside Road.

As I drove along the winding pavement, I was impressed to see verdant patches of green grapevines on my left, spreading down the valley floor toward the Russian River, which snakes its way slowly toward the Pacific Ocean. On the right side rose the lower hills and benchland of the coastal mountain range. Here, wineries peeked between large oak trees every few miles, interspersed with a few vineyard patches. Wild flowers appeared in bright spots of red, yellow, and orange along the roadside, and further up in the hills, tall fir trees rose in dark green patches.

Eventually a sign announcing the turnoff for John Tyler Winery appeared on the right, and I steered my car into a gravel driveway that led to a small tasting room tucked back in the trees. As it was ten o'clock in the morning, the only other car in the parking lot was an old chocolate-brown Mercedes. Beyond the car and seated at an outdoor table with a large sun umbrella, was a tiny silver-haired lady who I recognized immediately as Helen Bacigalupi. She and her husband Charles were the owners of the legendary vineyard that had produced a large percentage of the Chardonnay grapes that went into the Château Montelena wine that won the Judgment of Paris in 1976.

"Welcome! Come and join me." Helen stood up and gestured to me, and I could see that she had dressed her four-foot-ten-inch frame in a becoming burgundy-colored corduroy blazer, with navy slacks and sensible leather loafers to walk in the vineyard. She shook my hand with a strong grip and smiled, her blue eyes twinkling behind round glasses. She looked every inch the successful grape grower and smart businesswoman that she was known to be—even though she was eighty-seven years old.

"I hope you didn't mind meeting here at my son's winery," she said, as we settled comfortably into two chairs on the sunny deck of the tasting room.

"Not at all."

"It's just that it is easy to miss the turnoff for the Paris Tasting Block."

"Is that what you call it?" I asked. "The Paris Tasting Block?"

Helen smiled, and her face lit up. "Yes, ever since we heard the grapes from our vineyard won at the Judgment of Paris, we've called it that. You see," she continued, "we have over one hundred and eighty acres of vineyard here in the Russian River AVA, but the Paris Block is composed of four acres of Chardonnay vines planted in 1964."

"I can't wait to see it," I said eagerly, consumed with curiosity to view the famous vines.

"We'll drive over there shortly," said Helen, "but first I thought it might be more relaxing to sit here in the shade of this umbrella and talk a bit about the history of the place. Are you interested in that?"

"Yes, of course," I said and settled back in my chair to listen and enjoy the balmy weather of an 80°F spring day in the beautiful Russian River Valley.

History of Bacigalupi Vineyard

The Bacigalupi Vineyard is located in the central part of the Russian River Valley AVA and occupies what was known as the Goddard Ranch—an original homestead established in the early 1800s. The Goddards lived on the 121-acre property for over five generations before selling it to Helen and Charles Bacigalupi for $35,000 in 1956. The purchase included two old houses—one that was 100 years old, and the second that was 150 years old. The Bacigalupis moved into the 100-year-old house and slowly started updating it.

At the time, both Helen and Charles had full-time jobs in Healdsburg, where Helen worked as a pharmacist and Charles owned a dental practice. However, once they bought the property, Helen gave up her job to oversee agriculture operations.

"There were already some old vineyards on the property," reported Helen, "filled with field blends of Zinfandel, Alicante Bouschet, Golden Chasselas, and Mission grapes. In addition, we had acres and acres of prune orchards, but we couldn't make much money on prunes, so we decided to plant more vines."

The Bacigalupis consulted Bob Sisson, the local UC Davis grape advisor, who was also a dental patient of Charles's. Bob suggested they plant some of the newer grapes such as Chardonnay and Pinot Noir because they would perform well in the more moderate climate of the Russian River. However, neither Helen nor Charles had ever heard of those types of grapes and couldn't find a place to buy them. Finally someone suggested they contact Karl Wente in Livermore who had started a winery and nursery several years before. So Charles telephoned Karl who invited him to drive to Livermore, where Charles purchased the budwood for what has now become known as the famous Wente Chardonnay Clone. They grafted the Wente Clone to St. George rootstock with the assistance of Joe Rochiolo who lived down the road and had taken a class at UC Davis to learn how to graft vines.

Chardonnay vines in Paris Tasting Block

The resulting four acres of Chardonnay became known as the Paris Tasting Block when Mike Grgich, winemaker at Chateau Montelena at the time, contracted to buy the fourteen tons it produced in 1973. Mike blended it with other Sonoma and Napa grapes to produce the 1973 Chateau Montelena Chardonnay that won the 1976 Judgment of Paris, beating out wines from Puligny-Montrachet and Meursault Charmes. This single event caused people around the world to recognize that California was producing wines of exceptional quality.

Once the results of the famous tasting in France became known, Helen had many offers for her Chardonnay grapes and to this day continues to sell them at a premium to high-end wineries in Napa and Sonoma counties. The original Paris Tasting Block, planted in 1964, still exists and is surrounded by many other blocks of Chardonnay and Pinot Noir vines. The total Bacigalupi holdings now include forty-one acres of vineyards on what Helen and Charles still refer to as the Goddard Ranch. In addition, they have bought several other vineyards including Bloom

Ranch with seven acres of Chardonnay and Frost ranch with sixty acres of Pinot Noir, Zinfandel, Chardonnay, and Petite Sirah.

Ironically, many people are still not aware of the fact that 87 percent of the Chardonnay grapes that went into the 1973 Chateau Montelena, were from Sonoma County. However, the information is well documented in George Taber's book, *Judgment of Paris*:

> *"Just over 40 tons of Chardonnay grapes in 1973 were purchased from four suppliers: Charles Bacigalupi from Sonoma County's Russian River Valley; Lee Paschich, whose vineyard was located about a mile from the winery; Henry Dick from the Alexander Valley in Sonoma County; and John Hanna. Bacigalupi provided 14 tons, Paschich about one ton, Dick about 20 tons, and Hanna four tons. The Chateau Montelena Chardonnay was thus made with predominately Sonoma Valley grapes, although it was produced in the Napa Valley. (p. 143)"*

Touring Bacigalupi Vineyard

"Are you ready to go see the vineyard now?" Helen asked.

"Of course," I jumped to my feet.

"OK, then follow me in my car."

I saw Helen into her old Mercedes sedan before climbing into my vehicle and following closely behind. We didn't have to drive very far, because Helen turned onto Goddard Road just a few minutes after we drove north on Westside Road. We traveled on a narrow, paved road that climbed the hill and made several sharp turns before dead-ending in front of a small wooden house with several outbuildings and barns surrounding it. All around the buildings and climbing up the slopes were rows of vines on tall trellis systems.

"That's the one hundred-and-fifty-year-old house you're standing next to," said Helen as she climbed out of her car. "Look closely and you can see that the walls are made of redwood."

I stared at the old building, which looked small and dilapidated.

"No one lives in it," said Helen. "We just use it for storage. Instead, we live in the one-hundred-year-old house over there." She swung around and pointed at a larger house that looked more modern and was surrounded by gardens and trees. Just across the driveway was another vineyard.

"Wow, you live in the vineyards," I said, surprised.

"Yes, so when people tell me that spraying sulfur on vineyards is dangerous, I say nonsense. Do you think I would live where I could be poisoned?" She laughed and starting walking toward a block of vines near a large barn.

"Where is the Paris Tasting Block?" I called after her.

"Right here!" She continued to walk straight ahead and then turned around to motion me forward. "Here is one of the vines where Mike Grgich sampled the grapes that day in 1973 when he came to visit."

I followed rapidly behind her and then approached the vine with awe. Here was living history. This huge, magnificent grapevine was part of the story that put the California wine industry on the map. Its fruit had gone into a bottle of wine that French wine experts, in a blind tasting, had declared was better than some of the most famous wines of Burgundy.

Staring at the vine, I realized it stood about seven feet tall on a modified California sprawl trellis system, which reminded me of some of the taller pergola trellises in Italy. Healthy green leaves spread out in many directions, and the shoots were filled with many small grape clusters in full bloom with small white flowers. The trunk was massive, about ten inches across but also quite tall, so that the grape clusters were around shoulder height.

Helen walked over and stood next to the vine. It dwarfed her four-foot ten-inch frame. "So what do you think?" she asked.

"I'm in awe. I can't believe how tall these vines are."

"Yes, that is the way we like to plant them. We call it a modified California-sprawl trellis system. It works well for us and is easier to pick the grapes during harvest, because they are higher. The only problem with this block is that some of them are dying of old age now. See the missing spaces."

I glanced up the row of vines and noticed a few spaces where grapevines has formerly stood, but now all that was left were wooden stumps with tall golden grass growing around them.

Helen Bacigalupi in Paris Tasting Block

"My son John is trying to replant some of them using the same budwood," said Helen, "but it is very difficult to get the old ones out.

Their roots run very deep."

"That's wonderful that he is trying to replant with the same budwood," I said, still filled with wonderment over the beauty and height of the vines. I walked back over to the one where Helen said Mike Grgich had tasted the Chardonnay grapes.

"How did Mike hear about your vines?"

Helen grinned, and her blue eyes twinkled again. "You know to this day, I'm still not sure how he heard about us. We were selling the grapes to other wineries back then, including Chateau St. Jean, so he must have found us by talking to other people. Anyway he called me one day in August of 1973 and asked if he could see our Chardonnay."

"Yes," I encouraged, "so what did Mike say when he saw the vineyard?"

"Well, he walked around and looked at several different blocks and started tasting the fruit, but when he got to the Paris Tasting Block, he picked several grapes and after eating them said, 'Boy, these are the most beautiful grapes I ever saw in my life. The flavors! There are so many different flavors!'"

A delightful laugh slipped from her throat as she recounted the story. "So I negotiated the contract with him for eight hundred and fifteen dollars per ton, and then he came back almost every other day to walk through the vines and taste the grapes. Finally, he said they were ready to harvest at around twenty-three degrees Brix and asked me to deliver them to Chateau Montelena."

"So you drove the grapes to Chateau Montelena yourself?" I asked surprised.

"Yes, I had some people help me pick, but Charles was working in town, so we loaded the grapes into an old trailer, which I attached to the back of that Volkswagen van." She gestured to an old rusty Volkswagen parked in a wooden shed near the house.

"You did?" I asked in amazement, looking at the old van and thinking of the steep drive over the Mayacamas Mountains to Napa Valley pulling a trailer of grapes. "I know from the records that you sold them just over fourteen tons, so how many trips did you have to make?"

"It ended up taking five trips, and I took Highway 128 through Alexander Valley and down into the northern part of Napa Valley. That Volkswagen didn't have a lot of guts, though, and so I had to gun it to get it over the hill and hope no one was in front. Fortunately back then, there wasn't much traffic. Later, we got a used Chevrolet, and it had a lot of power. It actually 'growled' in low gear." She grinned widely.

"I'm impressed," I said and meant it. This eighty-seven-year-old woman was not only a brilliant grape negotiator, but she was also quite strong and gutsy. "So how did you hear that your grapes had won in Paris?"

"Mike called after the tasting in 1976 and asked if I had heard the good news," reported Helen. "When I said no, he told me how the Chateau Montelena Chardonnay had won and then asked if I would sell him more of the grapes. But I had to tell him no, because they were already under contract to be sold to someone else."

"And you've been selling them all of these years?"

"Yes, once people heard that these Chardonnay grapes were part of the winning blend, we've always had plenty of buyers and have been able to demand a premium price for them. Right now they are all going to a winery in Napa. Would you like to see some of the other blocks?"

I nodded, and she led me past the Paris Tasting Block to see some newer sections of Pinot Noir and Chardonnay, where the rows were more tightly spaced. The newer vines were obviously younger, because they had slimmer trunks, but they all appeared to be vibrantly healthy. They marched in straight lines over hills and down again, and I was impressed with how vast the Goddard Ranch Vineyard was.

Tall golden grass at least two feet high grew between the rows, and Helen explained that it was part of their sustainable farming program. "These are native California grasses," she said proudly. "We wait until it has completed setting its seed, then we mow it. If you mow too early, it will try to grow again and suck up all the water—competing with the vines. We've always let the wild grasses grow between the vines. I remember one time when some of the Fetzer boys came to visit—before they became organic—and they asked why we didn't plow everything under so only soil was showing, which was the custom at that time. We said we thought leaving the natural grasses was all right, and now it is back in fashion again to have a cover crop." She laughed, and her blue eyes sparkled in the bright sunlight.

As we walked through the rows, I asked Helen a question that had been on my mind. "So what does Bacigalupi mean?" I made sure to pronounce the "c" as a "ch" as I had heard her do.

"My husband Charles is of Italian ancestry, and as far as we know, 'baci' means 'kiss' and 'lupi' means 'wolf.' I don't know what the 'ga' stands for."

"Kiss of the wolf," I mused out loud. "That suggests a linkage to werewolves or to Romulus and Remus, the founders of Rome who were reared by a she-wolf."

Helen shrugged and then grinned. "Who knows? The name is from Charles's side of the family anyway."

Bacigalupi Vineyard Specifics

As we continued our tour, Helen provided additional information about the Paris Tasting Block and Goddard Ranch. Of the *forty-one acres* of vineyard at Goddard Ranch, fifteen are planted to Pinot Noir, and twenty-six are Chardonnay, including the four acres in the Paris Tasting Block. The **soil** consists of red volcanic clay with a strong iron content. It is also known as part of the Arbuckle Series, which are well-drained soils that formed in alluvial materials and are usually found on benchland or terraces.

The **elevation** is one hundred feet, and the vineyard is considered to be in Region III. *Average temperatures* are around 85°F in July with lows of 45°F at night.

"We usually have a forty-degree swing in temperature each day in the summer," reported Helen. "It keeps good acid levels in the grapes."

Rainfall matches the Sonoma County average of thirty-nine inches per year.

Rootstock is primarily St. George, but they also have AXR1. "We like St. George," said Helen. "It has worked for us, so why change? We did put some Pinot blocks on AXR1, and now we regret it because of the risk of *Phylloxera*."

Vineyard Specifics for Bacigalupi (Goddard Ranch)

Total Vineyard Acres	41 (4 acres of Chardonnay in the Paris Tasting Block)
Varietals	Chardonnay and Pinot Noir
Soil	Arbuckle Series—red volcanic clay on benchland
Elevation	100 feet above sea level
Average Temperature	Average July day temperature is 85°F, dropping to 45°F at night
Rootstocks	St. George, AXR1
Clones	Chardonnay: Wente; Pinot Noir: Martini Clone
Sun Exposure	Southeastern
Spacing	12 by 8 and 11 by 6
Trellis Systems	Modified California sprawl

Helen's son John, who has taken over much of the daily vineyard operations, provided information on the *clones*. "We have several different clones of Pinot Noir on the various ranches," John reported, "including Pommard, 777, and 667, but within the Goddard Ranch we are only using the Martini Clone for our Pinot Noir."

The Chardonnay is planted solely to the original Wente Clone, and whenever possible, John uses budwood from the Paris Tasting Block to graph over new vines.

Vineyard exposure in the Goddard Ranch is mainly southeastern, with the sun rising over the Mayacamas Mountains to the east and setting in the west over the coastal range behind the vineyards. *Spacing* in the Paris Tasting Block is twelve by eight feet, with the *trellis system* a modified California sprawl with cane pruning.

"We've modified it some so it doesn't sprawl as much," commented Helen, "but we still think this is a healthy trellis system, and we use it for all of our Pinot Noir and Chardonnay on this ranch. Spacing in the newer blocks is eleven by six."

Farming Practices—Sustainable

Helen and her family are using sustainable *farming methods* in all of their vineyards. This includes most of the recommendations outlined in the *California Code of Sustainable Winegrowing*. However, at this time, they have not pursued official certification. Farming methods for the Bacigalupis are based on personal judgment, experience, and frequent observation of the vines. The consistency in owning the same parcel for many years lends itself to this type of farming.

Fertilizer includes winter cover crop in the form of natural grasses that not only provide nutrients to the soil when it is cut but help prevent erosion and is a good habitat for beneficial insects such as ladybugs. Other fertilizer includes manure and compost made from the used grape skins (pomace) after winemaking. *Weed control* includes mowing between the rows and using herbicides such as Roundup under the vines.

"I don't believe in letting weeds grow up under the vines," Helen explained.

John explained that all of the vineyards are hand-tended by trained crews that use precise *canopy management* techniques such as removing excess leaves, shoots, and grape clusters.

228

"This increases air movement through the canopy," said John, "and better light penetration, drastically reducing the need for fungicides."

Helen with Chardonnay Grape Cluster in Bloom

In terms of *disease control*, they only use dry sulfur to prevent powdery mildew. Application is as needed and after precipitation or after any moisture gets in the vineyard. *Pest control* is not much of an issue because the property is deer-fenced. Furthermore, the installation of owl and bat boxes encourages those creatures to patrol for gophers, small birds, insects, and other predators that could harm the grapes. The use of cover crop in the vineyard also helps decrease dust, which naturally reduces dust mites that could harm grapevines. The biggest issue, according to Helen, appears to be wild hogs. "They can get in and strip the grapes right off the vines." Interestingly, this is a similar problem in the vineyards of Italy, France, and Germany.

Regarding *irrigation practices*, when they first bought Goddard

Ranch, Helen and Charles used only dry farming. However, they installed drip irrigation in the 1960s, and the Paris Tasting Block uses this method.

Farming Practices at Bacigalupi

Certifications	Not certified but following California Sustainable Winegrower methods
Fertilization	Green cover crop, natural composts of manure and grape skins
Weed Control	Mowing, Roundup
Canopy Management	Leaf, shoot, and cluster thinning as needed
Disease Control	Dry sulfur to prevent powdery mildew
Pest Control	Wild hogs are a minor issue.
Irrigation	Drip- and dry-farmed for old vines
Technology	No technology used in vineyard
Harvest Measurements	Generally 24°–26° Brix for Chardonnay and Pinot Noir

"I realized we needed to install irrigation," recalls Helen, "when we were selling Pinot Noir to Rodney Strong one year. I called him to say the Brix was twenty-three degrees, but the leaves were falling off. He said to harvest right away because if leaves were falling off, the vines were shutting down, and the fruit would not get any riper. We did, and it made me realize we needed irrigation." The source of the water is natural rainwater collected in two small ponds on the property.

At this time, the Bacigalupis are not using any *technology* in the Goddard Vineyard.

"Frost is not a problem here," reports Helen. "However, we do have frost protection equipment on the parcel next to the river." In addition there is no use of weather stations, water sensors, pressure bombs, petiole analysis, or other common technology. Instead, they rely on "grower instinct."

Regarding *harvest measurements*, Helen and her family discuss optimal picking dates as part of the grape-contract negotiation.

"I usually do one-year contracts," explained Helen, "and they specify that the buyer and seller work together to pick the harvest date." Generally, the winemaker or a designated representative visits in the weeks leading up to harvest to sample the grapes and measure brix, acid, tannin structure, and flavor development.

"Nowadays," said Helen, "it is usually around twenty-four to twenty-six degrees on Pinot and Chardonnay, but the sugars that went to Paris were twenty-three degrees. They picked earlier back then."

Vineyard Economics for Bacigalupi

As we were wrapping up our vineyard tour, a small four-wheel vehicle with three workers appeared out of the vines. As they passed us, they waved, and I could see they were dressed in the garb of vineyard workers with jeans, work boots, loose shirts with long sleeves, and white caps with neck flaps to protect them from the sun. They stopped near the barn and jumped off the vehicle to wash their hands at an outdoor water pump.

"Are they your vineyard workers?" I asked Helen.

She nodded. "Yes, they've probably come in for their lunch break. They are full-time workers who live here on the property. Two of them are brothers."

"Do you think they would mind if I took their picture?"

"You can try, but they are usually quite shy about photos."

I quickly walked toward the trio and greeted them in Spanish and then requested permission to take their photo. All three motioned "no" with their hands and walked quickly away.

"Told you they were shy," said Helen as I walked back over to where she was standing in the driveway. "They are good and loyal workers and have been with us for over thirteen years now."

"Do they help with the harvest as well?"

"Yes, and they recruit other workers to assist. There are also a couple of local women who are friends who come in to help with suckering and thinning the vines in the spring and will sort the grapes during harvest."

Helen and Charles on 4-wheeler

"Since we are on the subject of workers, can you share your wage rate?"

"We pay an average of $12 per hour for seasonal workers, but at harvest we pay by the bin, which is one thousand pounds. The workers pick in teams and divide the money. If they pick fast, they can earn up to fifty dollars per hour."

"That's great money!" I said in surprise. "So in looking at the economics of the vineyard—revenues and costs—how would you say you are doing?"

Economic Viability of Bacigalupi Vineyard

Average Yield	Approximately 2 to 4 tons per acre. Paris Tasting Block: 1 to 2 tons per acre depending on year
Total Average Tons Per Year	Average 123 tons in 41 acres of Goddard Vineyard
Costs Per Acre	Similar to average Sonoma County farming costs of $5,000–$6,000
Revenues	Above average Sonoma County prices of $2,235 per ton—negotiated by individual contract and block, e.g., Paris Tasting Block brings in $4,500–$5,000 per ton
Economic Health	Very good with strong list of clients

"Depends on the year," said Helen, "but in general, pretty well because we own all the land, and I don't believe in debt." She explained that negotiated price per ton depended on the block location, age, and varietal, but that they generally achieved higher prices than the Sonoma County average ($2,235 in 2013). The Paris Tasting Block achieves a premium price because of its historical significance and high quality, and negotiated prices usually range from $4,500 to $5,000 per ton. They currently sell to thirty-six different wineries, including such famous labels as Rudd, Williams Selyem, Armida, Arista, and Divino.

Average yield is two to four tons per acre, depending on the vintage and block. The Paris Tasting Block still achieves two tons in good years but can also drop to one ton per acre in poorer vintages. Therefore, the total average for all forty-one acres is three tons annually, resulting in a total *average yield* of 123 tons.

Detailed farming **costs per acre** could not be provided, but John estimated that it was most likely on par with the Sonoma County average of $5,000–$6,000 per acre annually, including labor costs.

Soul of the Vineyard—"Famous" and "Rooted"

As we walked back toward my car, Helen stopped to stand in the shade of a large tree near her one-hundred-year-old house, and I asked her to describe the vineyard in one word.

"I'm not sentimental about vines," said Helen, with a spark in her eye. "They are not people."

"I realize that, but if you could sum it up into one word, what would it be?"

Helen shrugged and then swung around to look over at the Paris Tasting Block. "OK, *famous*. I would have to say that it is nice that these grapes got an award. It helped California enormously, though at the time, I don't think we really appreciated it. However, they are just vines and not particularly any different from other blocks. They are not people," she repeated. "And if they are not producing enough, so long, they've got to go. We are still trying to make a living here."

I cringed at her words and thought what a shame it would be to tear out such legendary vines. "Well, tell me what you like best and least about being a grape farmer," I asked, trying to change the subject to something more positive.

"Well," said Helen squinting up at the sun and then moving her gaze to focus on a grove of redwood trees near the vineyard block across the driveway. "We love living here. See those redwoods there." She gestured to the tall red-barked trees. "Charles and I planted those for his mother because she loved redwoods. It has been wonderful to live in the country

234

surrounded by nature and be filled with peace and quiet. It is fulfilling to leave this legacy to our family. Our son John really loves this place, and now his wife and daughters are helping to run it. We have a new great-granddaughter, too. Perhaps she will be involved someday."

Redwoods Near Vineyard

As she said these words, a warm smile lit up her face, and I hated to disrupt it by asking her about the challenges of farming.

"For me, coordinating with all of the buyers can often be quite challenging," she answered smartly. "I have to allocate who gets which blocks and even rows of vines. In some cases we have two to three different buyers for one block. They each want a certain tonnage, but some years there is more or less production, and I'm always worried about excess and shortages. My daughter-in-law is managing all of this now, but it is an incredible amount of work." She sighed and looked around the vast blocks of vineyards. "Scheduling picking can also be difficult. You have to pick at the optimal time, but sometimes I can't get the workers scheduled then, or the wineries aren't ready because their tanks are full."

"It sounds challenging. What would you say working here has taught you?"

Bacigalupi Family

Helen's gaze wandered over the rows of tall vines across the road. "I've learned that farming is a lot of work, especially when you do it yourself, but if you want quality, you have to do it yourself. Over the years, so much has changed on this property." She waved her hand in a sweeping gesture encompassing all of the property.

"We've planted all of these vines ourselves and had to clear out the old orchards and vineyards to do so." She paused, and her voice became softer as she continued. "And I imagine the original settlers had to clear all of this property in order to plant the old Zinfandel vines that are now gone—replaced by our Chardonnay and Pinot vines. And they did it with horses! We found the old horse bridles in the barn over there with the tractors." Helen turned and pointed at the barn.

"But this is a wonderful legacy to leave our son and his family," she said with emotion in her voice. "He loves the land like my husband. They are both sentimental about the vines, but I'm not. This is still a business. Charles's mother tried to stop him from buying this land. She told him that farmers only have rocks in their pockets and no money. But one thing we were both in agreement on when we married was that we wanted to buy land, have a home, and grow grapes. We love it here."

Later I followed up with Helen's son, John, to ask him to describe the vineyard. Interestingly his response echoed the sentiment of his mother's in regard to family and legacy, but his connection to the land rang through clear and strong in his one-word description of "rooted."

"I would have to describe this vineyard as *rooted*," he said, "and by this I mean from the deep roots of these vines to the deep roots of our family of four generations, we are blessed to call this home."

As I analyzed my experience of the Bacigalupi Vineyard and its famous Paris Tasting Block, I continued to feel a strong sense of awe at the thought of those tall Chardonnay vines that produced the rich fruit that helped to change the course of wine history. I think of the Bacigalupi family who has tended the vines for four generations and the Goddards who homesteaded and lived on the property for five generations before— all of them dedicated to the land, sending deep roots of commitment and hard work into the soil to craft a product in partnership with Mother Nature. And I think of the old gnarled trunks in the Paris Tasting Block that have died and are covered with tall golden grasses.

"It is very difficult to get the old ones out," Helen's voice echoes in my head. "Their roots run very deep." But I know that someday, John will find a way to gently uproot them, so he can graft the parent budwood to new rootstock in order to create the next generation of deeply rooted vines.

Signature Wine: Edge Hill Bacigalupi Vineyard Chardonnay

Over the years Helen and Charles sold the Bacigalupi Chardonnay grapes to a number of different wineries, but for the past decade or more, the main buyer has been Rudd Winery in Napa Valley. Sold under Leslie Rudd's Edge Hill label, the Bacigalupi Chardonnay has scored consistently high with wine critics over the past several years.

I was fortunate enough to taste the 2011 Edge Hill Bacigalupi Vineyard Chardonnay with winemaker Patrick Sullivan. A Sonoma County native, Patrick has an impressive wine pedigree including a Masters in Viticulture and Enology from CSU Fresno and work experience in such prestigious wineries as Peter Michael, Paul Hobbs, and Lewis Cellars. He worked at Rudd for more than seven years and has recently taken a new position with Spring Mountain Vineyard. Tall and lean with short dark hair and olive-green eyes, Patrick has a direct style of communication, a good sense of humor, and enjoys being outdoors in vineyards just as much as he relishes making wine in the cellar.

"I remember the harvest of 2011," said Patrick. "It was a cooler year than normal, and we didn't pick the Baci Chardonnay until the third week of September. I recall being there in the vineyard to help pull leaves and pick grapes with the harvest crew, Helen's son John, and her granddaughters."

"You actually helped harvest the grapes in this bottle?" I asked in surprise.

"Yes," smiled Patrick. "I always assist in the harvest. It is part of the job that I really enjoy."

"So you know the Bacigalupi family rather well?"

"I've spent many hours discussing viticulture issues with them in

Helen's kitchen. It's like being with my own family. We had many discussions and even some bickering, but it is all part of the warmth and enjoyment."

"So what were your thoughts of the Bacigalupi Vineyard the first time you saw it?"

"Old school," responded Patrick, "and by that I mean it was like a vineyard from a different century. The grass between the rows is very tall, and the vines are allowed to sprawl and grow naturally without much interference. Of course, they do some canopy management such as pulling leaves to allow the air to circulate on the grapes, but otherwise it is a 'very natural' vineyard."

"So why do you think the vineyard produces such exceptional grapes?"

"I believe it is a combination of factors. Primarily it is a great location for vines. The soil, site, and climate are perfect for Chardonnay. Another factor, I believe, is that there is a great mix of clones in the vineyard."

"But it is the Wente Clone," I said, puzzled by his statement.

"Yes, but keep in the mind the Wente Clone is actually a combination of clones as well. Do you remember how Mike Grgich kept exclaiming about the flavors in this vineyard? Well, when I walked around and tasted the grapes, I could tell it was an old-field blend of Wente clones. Even today, if I was blindfolded, I could taste the difference between some of the vines. That is the magic of the Bacigalupi Vineyard. It has so many wonderful and different flavors of Chardonnay."

"Interesting," I said. "That reminds me of many of the vineyards I visited in Burgundy. When I asked the winemakers which clones were in the vineyard, they just shrugged their shoulders and replied 'mixed selection,' meaning it was a combination of different clones that had developed over the years."

Patrick nodded enthusiastically. "Yes, in a way, it is like a vineyard orchestra, and I think it is what makes the Baci Chardonnay so special—that and the fact that the vines are older."

"So you can taste the difference between the Baci Chardonnay and others?"

"Definitely. We had many blind tastings at Rudd, and the Baci Chardonnay always stood out as distinctive and special."

That was a signal for me to try the wine, so I picked up my glass and immediately noticed the warm golden color, which is a sign of a Chardonnay that has been aged in oak for some time. Swirling the wine in the glass and bringing it close to my nose, I was immediately enveloped in a cloud of lemon and golden apple with a hint of minerality. Taking a sip, I swished the wine around in my mouth for a long time and was surprised at the many complex flavors. The rich lemon notes continued and were joined by pineapple, subtle butter, well-integrated toasty oak, and an exquisite acidity that made my mouth tingle in a delightful way. The finish was extremely long and a little warm due to the 14.2 percent alcohol.

"So what does this wine taste like to you?" I asked Patrick.

"It tastes like Baci."

"What do you mean, 'it tastes like Baci'? And do you always used the shortened term 'Baci' for Bacigalupi?"

Patrick laughed. "It is much easier to say 'Baci." Then he paused and seemed to be searching for words. "I always have a hard time explaining this," he said, "but Baci Chardonnay to me has an uplift to it. Obviously it has citrus, minerality, and toasty notes, but it is that uplift at the finish that makes it special. It has both weight and an ephemeral component to it that makes it Baci Chardonnay. I think it is the vineyard speaking in the glass."

"So do you try to allow the vineyard to express itself in your winemaking?"

"Yes," he said immediately. "I try to find out how the wine wants to be made, and it may be different depending on the vintage, but with Chardonnay it is all about élevage—or aging."

"So tell me how you made this."

Patrick described that after he helped harvest the grapes at around 24° Brix, they were loaded into refrigerated trucks and transported to Rudd Winery where they were immediately whole-cluster pressed and then allowed to settle in tank for one day. If necessary, adjustments were made at that time, such as adding nutrients, water, or acid, but he did not use SO_2 until after malolactic fermentation. The juice was then transported to 225-liter French oak barrels that were 40–50 percent new. Alcoholic fermentation was allowed to start naturally with no added commercial yeast in a 60°F cellar. In some cases, the wine took up to two weeks before it started to ferment.

Once alcoholic fermentation was completed and malolactic fermentation (ML) began naturally in barrel, he conducted *battonage* every two weeks, slowing to once a month as ML finished, usually in March. *Battonage* is a French term that refers to stirring the wine in the barrel in order to mix the lees (particles) on the bottom into the wine. This process usually adds a nutty quality to the wine as well as additional depth and creaminess. It is a method used by most Chardonnay winemakers in Burgundy and many in California.

Patrick didn't add SO_2 until ML was complete. The wine was topped as needed, but never racked off the gross lees until it was ready for bottling after fifteen months of oak aging. The French refer to this as "*sur lie*," meaning the wine rests on the lees until ready to be bottled.

Patrick said the wine was bottled without fining or filtering.

"I like to make my wines as naturally as possible," explained Patrick. "I don't use very much SO_2 in the beginning because I want the Chardonnay juice to brown out."

"Brown out?"

"Yes, like an apple turns brown naturally. By letting the wine brown out naturally, you can reduce bitterness as well as the total amount of SO_2 you add to the wine."

"Interesting," I said, very impressed with the depth of flavor and new elements I continued to find in the wine as I sipped it. "It is a truly magnificent wine."

"Thank you," said Patrick modestly. "I like to think that I helped to capture the magic of the Baci Vineyard. It is rare to find an old Chardonnay vineyard like that."

"You believe that old vines provide higher levels of quality?"

"Usually," said Patrick, "and that is why I'm encouraging my parents to plant more vineyards. When it comes to vines, I believe in the old Chinese proverb that goes something like: 'The best day to plant a tree is yesterday. The second-best day is today.'"

Chapter Ten

Hanzell Vineyards
Sonoma Valley AVA, Sonoma County

June is one of the most beautiful months in Sonoma County with the vineyards in full green leaf and the grape clusters growing larger each day. The fear of frost is over, and the majority of days are bright and sunny with balmy temperatures in the low eighties. During most evenings, the cooling fog rolls in from the Pacific to wrap itself among the vines, preserving acidity and flavors in the grapes.

It was just such a fresh June morning as I drove to Hanzell Vineyards, located in the lower foothills of the Mayacamas Mountains behind the town of Sonoma. The approach to Hanzell is rather convoluted in that I had to drive through several residential neighborhoods before coming to a narrow road that meanders through a field and then twists its way up a hill through sprawling California oak trees to an impressive gate. Hanzell is a by-appointment winery, so I leaned out the window to address an intercom near the entry, and after a quick exchange to confirm my appointment, the black, wrought-iron gate slowly opened. As I continued to drive up the hill, the first swath of lush, green vineyards appeared on my left, and then the imposing wood and stone tasting room, modeled after the twelfth-century press house at Clos Vougeot in Burgundy, rose up in front of me, perfectly situated above another sweeping vineyard block.

As soon as I entered the business office, I was greeted by the Director of Vineyard Operations, Jose Ramos Esquivel, and winemaker Michael McNeill. Jose wore a baseball cap over his thick white hair and had a

243

welcoming smile in a face that was tanned from long, sunny days working in the Hanzell Vineyards. Originally from Mexico, Jose had worked his way up to director of vineyard operations since first coming to Hanzell in 1975. His many years of experience working the land made him familiar with almost every individual vine on the forty-three acres.

"I see you brought a hat," said Jose, pointing to the large straw sunhat I carried in my hands.

"Yes, I thought I would need it on such a bright day touring through the vineyards."

"Good idea," smiled Jose, and reached out to shake my hand in a warm clasp.

Ambassador's Vineyard with Chardonnay Vines

"We thought you would like to start in the oldest section of the vineyard," said Michael, as he leaned forward to shake my hand as well. "It is called the Ambassador's Vineyard and was planted in 1953. It is located just outside next to the tasting room."

I nodded in agreement and followed them outside to a group of beautiful old gnarled vines set on terraced soil with generous spacing of twelve by eight feet. The vines were head-pruned and had healthy green leaves and grape clusters, which looked like small green olives at this time of the year. Directly behind the vineyard stood the historic winery building that had been turned into a tasting room several years ago.

Michael turned and gestured to the building. "As you probably know, this is the old winery building, which is based on the press house at Clos de Vougeot and is a two-thirds scale of the original structure." Michael was tall and thin with penetrating, blue eyes and a serious air but with an underlying passion for viticulture and enology. I knew he had an impressive pedigree as a winemaker with stints in Burgundy and Oregon, as well as time spent working at Chalone Vineyards.

"It was built in 1957," he continued. "Do you know the history of the estate?"

History of Hanzell

It was while working in Europe in the 1940s and helping to implement the Marshall Plan for President Truman that US Ambassador James Zellerbach first fell in love with wine. His main passion was for the wines of Burgundy, exquisite Pinot Noirs and Chardonnays, which he dreamed some day of creating in his homeland of California. So in 1948, he purchased the first acres of what would eventually become the two-hundred-acre property of Hanzell Vineyards, set in the foothills of the Mayacamas Mountains behind the town of Sonoma.

James decided to christen his property "Hanzell," a name that was created when he joined his wife's name, Hana, to the first four letters of his family name, Zellerbach. In 1953 he was able to launch his dream by planting four acres of Pinot Noir and Chardonnay in what has become known as the Ambassador's Vineyard. Today those vines are considered to be the oldest continually producing Pinot Noir and Chardonnay vines

245

in America.

Then in 1957 he built Hanzell Winery and modeled it on the press house at Clos de Vougeot. As a member of the Chevaliers du Tastevin, he had been indoctrinated at that prestigious Burgundian winery and wanted to bring the memory of it to Sonoma County. He hired Napa Viticulturist, Ivan Schoch, to oversee the vines and Brad Webb as winemaker. Together, the three of them started crafting beautiful wines and developed many innovations such as the use of custom-designed, temperature-controlled, stainless-steel fermentation tanks. They also developed the Hanzell Chardonnay Clone, were the first to use inert gas at bottling to prevent oxidation, and were the first winery in California to exclusively use French oak barrels for aging.

James died in 1963, three years before Robert Mondavi started his winery in Napa. He left behind a legacy of terroir-driven wines that achieved his vision to be "equal to the finest in the world." Today, according to Eric Asimov, *NY Times* wine critic:

> *"Hanzell has produced wines of subtlety and power that age beautifully and, above all, speak clearly of the vineyards (p. 2)."*

After James's death, the estate was sold to the Day family in 1965 before it was acquired in 1975 by the current owners, the De Brye family. During that time, the vineyard holdings on the two-hundred-acre property have been expanded so that today Hanzell Vineyards totals forty-three acres, of which thirty-one are planted to Chardonnay and twelve to Pinot Noir.

Touring Hanzell Vineyards

"So is this vineyard comprised of the famous Hanzell Chardonnay Clone?" I asked as we stood next to a three-foot-tall bush vine in the *Ambassador's Vineyard*.

"Yes," said Michael. "The original budwood came from Stonyhill Winery in Napa, which was started in 1947, but over the years, it has transformed itself into the Hanzell Clone."

Jose beckoned me to come closer as he gently cupped a cluster in his hand. "The Hanzell Clone is known for both big and small berries," he said, pointing at the cluster of baby green grapes in his palm. I noticed that some were the size of a small pea, whereas others were larger.

"We also call it hen-and-chick syndrome," Michael said, "but the French refer to it as *millerandage*. The clone is known for its floral notes, high phenolics, good structure, and crisp acidity. Currently about eighty-five to ninety percent of the vineyard is Hanzell Clone with the predominant rootstock of St. George. We also have a Hanzell Pinot Noir Clone, but that budwood is originally from Mt. Eden Winery."

I gazed out across the vineyard with its green leaves undulating in the slight morning breeze and was humbled to know that these were the oldest Chardonnay vines in America. The sixty-year-old vines were evenly spaced on slightly gravelly soil that fell in gentle terraces down the hillside. They seemed to stand proud and resolute above the newer blocks that were planted further down the slope.

"Within the Hanzell property," said Michael, "we have six subvineyards. To see the others we need to take the truck. Are you ready for a drive?"

I nodded yes, and we walked over to where a long tan SUV was parked. Michael gestured me into the front seat while he took the wheel, and Jose climbed into the back seat. Once buckled in, we drove down the hill to the *Zellerbach Vineyard*, which is distinguished by a tall grove of Douglas fir trees planted in the middle of the vines.

"We wanted to keep the trees because they are so beautiful," explained Jose, "and they are also a nice place for the workers to rest in the shade or have a picnic." I turned to look at him, and he flashed a quick smile and continued. "We also have some red-tailed hawks that nest in the trees and hunt gophers in the vineyards. This is a good thing."

I nodded in agreement, knowing from my own hobby vineyard experience how helpful hawks and owls are in patrolling the vines for gophers that can eat the roots of young vines. All types of wildlife are attracted to vineyards, and they often seem to work in a symbiotic system of checks and balances, with one form of wildlife serving as prey for another.

Michael parked the SUV, and we climbed out to walk into the Zellerbach Vineyard. "This is one of the largest blocks on the estate," said Michael, "with twelve acres of Chardonnay made up of six different clones and planted on different rootstock."

Zellerbach Vineyard

From where we were standing, I could look out on a sea of waving, green Chardonnay leaves and felt excited about the great wine that would come from these vines in the future. Turning around, I looked back up the hill and could see the Clos de Vougeot building towering over the Ambassador's Vineyard above and realized that the two vineyards were connected.

"This vineyard was planted in 1998," said Michael, "but the soil is the same in both vineyards."

Looking down at the ground, I saw the dirt was gray-brown in color with small rocks. "What type of soil is this?" I asked.

"Raynor-Montara," responded Michael immediately. "It is a mix of clay and loam with small rocks and pebbles. It is well-draining, which is good for wine grapes."

"The Day Vineyard is the same," Jose contributed and nodded down the hill. "Do you want to see?"

I nodded yes, and we started walking down a dirt road to another block of Chardonnay vines. As we strolled, a red-tailed hawk circled above us, and I watched it swoop gracefully over the vineyard; its white-feathered stomach with gray-and-black stripes and its red tail feathers were outlined clearly against the vivid blue of the sky.

Jose walked into a row of vines in the *Day Vineyard*, and Michael and I followed. Here the vines appeared to be younger and were positioned on a VSP trellis so they rose about six feet on each side. The leaves were green with strong veins and a beautiful lacy edge, and the pale green grape clusters appeared healthy and growing each day in the warm sunshine.

"We have four point five acres here," said Jose. "Some of these vines were planted in 1972." His face was full of pride.

"Jose knows every vine on the estate," Michael said. "In fact, I think he sees them as his children."

"*Si!*" Jose grinned.

Back in the truck, we drove on the paved road for a short time before cutting off through a grove of oak trees and then turning back onto a dirt track. The truck climbed a steep hill, and then Michael pulled over and stopped the motor.

"Do you want to see the best view on the estate?"

"Sure!"

Jose led the way through a grove of tall eucalyptus trees, and then we were standing on the edge of a steep hill overlooking the *Ramos Vineyard*. The view was stunning, with the vineyard marching downhill in vertical rows toward groves of massive old oak trees beyond. In the distance Sonoma Valley stretched out before us in a patchwork of yellow grasslands, green vineyards, tall blue-green fir trees and bountiful oak groves with tiny houses and wineries. Rising up behind was the massive girth of Sonoma Mountain, home to steep forests and Jack London's historic ranch.

"The Ramos Vineyard?" I asked. "Any relationship to you, Jose?"

He nodded shyly and looked off into the distance.

"It was named for Jose," Michael interjected. "He has worked for so many years on this land that the current executives, Jean and Bob Sessions, decided to name this vineyard after Jose."

Jose Ramos in Hanzell Vineyards

"Bob and I planted this vineyard," Jose said, finally turning around to look at me, and I could see how much it meant to him. "But it is a worry to me at times." He pulled off his baseball cap and scratched his head.

"Why is that? It looks beautiful."

"Balance is the main worry in this vineyard," said Jose, gazing down at the neat rows with their long green canes. "It has incredible vigor and grows very fast. We always have to thin the shoots and leaves so the fruit can ripen."

"What clone is it?" I asked.

"It's the Hanzell Clone planted on St. George rootstock," Michael responded, "the same as the Ambassador's Vineyard, but for some reason it produces a lot of canes and leaves; however, the fruit quality is very good. It is possible that the high vigor in the vineyard is due to the different soil here—Red Hill Series clay. Also, this vineyard is located higher on the hillside."

Jose nodded. "It is above the morning fog and gets much sun in the afternoon."

"The *De Brye Vineyard* just across the road where we parked is even hotter," Michael said as he turned to start back toward the truck. But I wasn't ready to go yet and lingered for a few more minutes looking at the magnificent view over the Ramos Vineyard. It was the type of place that I would enjoy hiking up to and sitting there for hours just gazing out across the valley and dreaming.

As soon as we walked across the road and into the De Brye Vineyard, I could feel the sun shining more warmly on my face. The rows spread horizontally across the face of the mountain with a clear southern exposure and distant views of the San Pablo Bay and San Francisco. It was the largest block on the estate at fifteen acres, and included both Pinot Noir and Chardonnay. We walked down a row, and I immediately noticed the different trellis system.

"Is this a new type of trellis and pruning system?"

Michael laughed. "Yes, you could say so. These Pinot Noir vines used to be head-pruned, but we modified them into cane-pruned vines and have added a T-bar so the top wire divides the shoots to form a shade canopy. It is similar to a *Y* or shorter Lyre system."

I reached out my hand to touch a fat shaggy trunk of one of the vines, and then traced one of the slender canes that were trained upward along a wire, so that the leaves rained gently down on both sides, creating a pretty curtain.

Jose leaned forward and lifted one of the long, leaf-laden shoots to point at the protected clusters underneath. "See," he said, "now the fruit doesn't get sunburned."

Valley View from Hanzell Vineyards

"Yes," said Michael. "We find this new system provides the perfect dappled sunshine that the grapes prefer, especially in this sunnier section of the estate."

"Wow, these are pretty plump," I said, and reached out to hold a Pinot Noir cluster in my hands. The grapes were still pale green but fat and tightly packed together, and I noticed some of the berries were slightly red at the tips.

Michael smiled. "Yes, they are ripening fast. In fact, we've had so much sun and warmth in the past month that the red you see on the edges of some of the berries is because they grew so fast it cuts off nutrients to those few outer berries."

"Interesting. I thought it was early verasion," I said, referring to the change in color in red-grape varietals that usually occurs in July and signals that the grapes are maturing for harvest.

"You can see the big difference that the location and exposure have on the ripening process," Michael stated. "Also, the soil here is more of the Red Hill Series clay, which is a combination of sandy loam and clay. It is well-draining and usually shows some red in color."

Jose bent down to scoop up a handful of soil and let it run through his fingers and fall gently back on the ground. It was a beautiful, rich brown with hints of red.

"Lovely dirt," I said, "and like the soil in Burgundy, France, it contains clay but not the limestone that region is known for?"

"No," responded Michael. "We really don't have limestone here, but we do have the clay component of Burgundy, which combined with the sandy loam, seems to work just as well."

Back in the truck, the next stop was the *Sessions Vineyard*, which was located on the far southeast side of the estate on a steep, rocky hillside. It took a while to wind our way down the hill from the De Brye Block and around to the Sessions Vineyard. The first thing I noticed was how varied the different sections looked in this four-acre section of Hanzell.

"As you can probably guess," said Michael, carefully keeping his eyes on the narrow and steep roads as we drove around the vines, "the Sessions Vineyard was named for Bob Sessions, and he decided to use this as an experimental block. Therefore, there are eleven subblocks, and each is planted with a unique rootstock and clone of Pinot Noir. It also has Red Hill soil and is our steepest location with a slope of thirty-eight percent."

"Many rocks, too," interjected Jose from the back seat.

"Jose should know," Michael laughed, "as he and his crew helped Bob dislodge many boulders and rocks so they could install this vineyard."

I was impressed with the innovation exhibited by planting such a varied block of Pinot Noir and was curious about how it played out in the winery.

"We are fermenting the subblocks separately," Michael answered my question. "It is fascinating to see how different they taste based on clone, but then again, Pinot Noir is famous for having so many unique-tasting clones. It allows us to blend some amazing wines."

Hanzell Vineyard Specifics

As we made our way through the various blocks of Hanzell Vineyards, I asked Jose and Michael to provide some technical details.

"Since the forty-three acres are dedicated to Chardonnay and Pinot Noir," said Michael, "the *climate* is obviously cooler. Fog and wind sweep through Stage Gulch Pass." He pointed to a dip in the mountain range to the southwest. "This helps to cool down the grapes in the hotter afternoons. We range from the mideighties in the summer months to midfifties at night."

In terms of *soil*, Hanzell Vineyards has a mixture of Renard Montoya Complex and Red Hill Series clay. The *elevation* ranges from five hundred feet in the Day Vineyard up to eight hundred at the top of De Brye. *Row orientation* is varied due to the slope of the land and sun exposure.

"We have been very careful in laying out the vineyard so the vines receive the best sunlight for ripening but not too much so as to prevent sunburn," reported Jose. "Since we are right at the fog line, many mornings we wake up to fog and then have sunshine in the afternoons."

Hanzell Vineyards includes a diverse combination of *rootstock and clones*. The oldest vines from 1953 are on St. George rootstock, but the newer plantings include 3309, 101-14, 420A, and 1103P. Chardonnay clones comprise the Hanzell Clone, Robert Young, and the Wente Clone. Pinot Noir clones are Dijon 76, 667, 777, Pommard, 115, Wadensvil,

Swan, and Hanzell.

"We have noticed that the rootstock has a huge impact on ripening for Pinot Noir," stated Michael, "whereas the different types of clones impact the taste of the wine and make our blends quite interesting."

Vineyard Specifics for Hanzell

Total Vineyard Acres	43
Varietals	Chardonnay and Pinot Noir
Soil	Red Hill Series clay and Renard Montoya Complex
Elevation	500 to 800 feet above sea level
Average Temperature	Summer mid80°F during the day to mid-50°F at night
Rootstocks	St. George, 3309, 101-14, 420A, 1103P
Clones	Chardonnay: Hanzell, Robert Young, and Wente Pinot Noir: Dijon 76, 667, 777, Pommard, 115, Wadensvil, Swan, and Hanzell
Sun Exposure	Varies by block, but primarily west and southwest
Spacing	Old vines: 12 x 8 and 15 x 8; New plantings: 8 x 6 and 10 x 8
Trellis Systems	Bush vine, VSP, Y, or modified T-bar

In terms of *trellis systems and spacing*, the original 1953 Chardonnay vineyard is still on head-pruned vines with twelve by eight and fifteen by eight feet spacing. The newer blocks are on a VSP trellis system with eight by six or ten by eight feet spacing.

"For the most part, we use cane pruning because it produces more fruit," reported Michael, "although there are a few places where we use cordon, which is usually easier to prune and less expensive from a labor viewpoint."

Farming Practices—Sustainable

"We practice sustainable farming methods," Michael explained, "but are not pursuing certification at this time."

For *fertilization*, Hanzell uses natural compost in the autumn and plants cover crops each spring, which are tilled back into the soil as a natural fertilizer. They perform petiole analysis every year to check for nutrients.

"We often find we are low on phosphorus here, "said Michael "so we may add that to the soil. For *weed control*, we use Roundup as well as mow and disc."

In terms of *canopy management*, they have been moving to smaller canopies and will hedge twice each season. Hanzell will also drop fruit—perform a green harvest—if they think it is necessary and occasionally will remove extra leaves to provide more light to the growing clusters.

"Sometimes we pull leaves on the east and north sides of the vines," said Jose, "to make sure the grapes have enough sun."

"Yes," said Michael, "it is the southwest sun that is hottest here, so we usually allow leaves to remain on those sides of the vine to protect the clusters from sunburn. Some people think that this area is too hot for Chardonnay and Pinot Noir, but the winery has been producing both varieties here for over fifty years, so I don't think this is true."

For *disease control*, such as powdery mildew, they use a combination of organic and nonorganic sprays every ten to fourteen days when it is a threat but do not use sulfur. In terms of *pest control*, they have had to make modifications to deal with deer and birds.

"We've had to raise our deer fence from seven to ten feet, because the deer were jumping the seven-foot-high fence," reported Michael. "And we have two hundred acres of property with deer fencing, so it was a big job." Gophers are also an issue in the vineyard, and turkey and other types of birds can be a problem. "We have to net every year to keep the birds from eating the crop," said Jose.

Farming Practices at Hanzell

Certifications	None at this time but practicing sustainable farming
Fertilization	Natural compost, cover crops, additions when needed
Weed Control	Mowing, disking, Roundup
Canopy Management	Hedging and some leaf and cluster thinning
Disease Control	Organic and nonorganic sprays as needed
Pest Control	Increased height of deer fencing; must net each summer to keep birds from eating grapes
Irrigation	Drip irrigation, including old vines if needed
Technology	Weather stations, transmitter, petiole analysis, pressure bombs
Harvest Measurements	Brix: 23° or 24°; pH: 3.2 for Chardonnay and 3.4–3.5 for Pinot Noir. Ripe phenolics

For *irrigation*, the original vineyard was dry-farmed, but they have since added drip irrigation. "We don't use it unless needed," said Jose, "but new vines need more water." They have one well on the property that is thirty feet deep.

Vineyard *technology* is used at Hanzell to some extent. "During the

257

spring and summer we do a pressure bomb reading to see if any blocks or rows need water," explains Michael. "We also have two weather stations and one transmitter. We do petiole analysis once a year and will bring in consultants if we need additional information on something."

In terms of *harvest measurements*, as the winemaker, Michael had a ready answer. "We prefer to pick at 23 or 24 brix but always want to make sure we have good phenolic ripeness first. Acidity and pH are obviously different each year, but we prefer a 3.2 pH on our Chardonnay and a 3. 4 to 3.5 pH with our Pinots. The final alcohol level on both wines is generally 14% to 14.1%."

Economics of the Vineyard

When asked about the economics of the vineyard, both Jose and Michael were easily able to provide clear answers.

"Our *tonnage* is around 2.4 per acre for Chardonnay and 2.2 tons per acre for Pinot Noir," Michael reported. "We use most of the fruit to produce our own 6,000 cases but will sell a little as well in certain years." Therefore, the production for all forty-three acres would average around 99 tons per year.

"For the crew," said Jose, "we have two full-time people, but at harvest we will hire anywhere from 15 to 18 more workers."

Hanzell outsources seasonal labor to a vineyard management company and currently pays $18.50 per hour, which includes benefits.

"However, for the actual harvest," explained Michael, "the crews prefer to be paid by the tons, so we pay around two hundred and twenty-five dollars per ton."

"We pick at night," said Jose, "and have the workers dump into half-ton bins, which are transported by tractor to the sorting table."

Altogether, the average annual *farming costs* range between $5,500 and $6,000 per acre.

Exact *revenues* are difficult to calculate, as average FOB pricing cannot be revealed. (FOB means freight on board and is the price the winery sells to distributors, which includes costs of goods sold, plus profit). However, the suggested retail price per bottle of Hanzell is quite healthy and ranges from $36 to $95. They do have a three-tier wine club for private clients, and the remainder of the wine is sold via allocation—meaning the number of bottles sold each year is limited—to distributors and direct to trade where allowed.

Economic Viability of Hanzell Vineyard

Average Yield	2.2–2.4 tons per acre
Total Average Tons Per Year	99
Costs Per Acre	$5,500–$6,000
Revenues	6,000 cases, with suggested retail prices of $36 to $95 per bottle
Economic Health	Excellent; allocated list of customers to purchase wine

The Soul of the Vineyard—"Diversity"

Michael drove the truck slowly up the hill toward the tasting room and the historic Ambassador Vineyard. As we climbed out, I realized the day was even hotter now, and thought how good this weather was for the ripening grapes. We walked over to the Ambassador's Vineyard to stand near the oldest Chardonnay vines in North America, and as we all gazed at the beautiful vista of the green and healthy vines basking in the sun, I asked Jose and Michael if they could describe the personality of Hanzell Vineyards in one word.

Michael paused for a minute, reflecting upon his response before answering. "I think it would have to be *diversity,* because each vineyard has a distinct personality, and that diversity is based on the location of the block within the property and its distinct soil type, sun exposure, rootstock, clones, and many other factors."

Jose's answer supported Michael's viewpoint on diversity and at the same time was very poetic. "All the vines are close to my heart," Jose said, and pointed to his heart. I felt a lump come into my throat at his words, because the emotion he felt for the vineyard was very palpable.

Next, I asked them to describe the biggest challenges and joys in working the vineyard.

"I think the main challenge has been changing the pruning in different blocks," said Michael.

Hanzell Tasting Toom Modeled on Clos de Vougeot

Jose nodded his head in agreement. "Yes, it takes a while for the vine to adjust. Big cuts to the wood can be risky." Suddenly his face broke out in a shy smile. "But the best part is seeing the vines grow from year to year. It is especially joyful to see new vines I have planted grow healthy and reach maturity. You know, they are like children—you need to help them grow."

"For me," said Michael, "the best part of working here is seeing the connection of the vineyard to the wine. I can see how what the vineyard gives us each year is reflected in the wine. I also enjoy working as a team with the other employees and the crew during harvest and crush."

When I asked what working in the vineyard had taught them, they both described how they have learned to respect and accept what nature brings in terms of changes in weather, heat, and harvest conditions.

"I've come to realize that the changing conditions each year are really reflected in the vintage and taste of the wine," said Michael. "And Jose here," he patted Jose on the back with affection and pride apparent in his voice, "he seems to have an intuitive connection with the vineyard. He knows when it needs water, or not...and will tell us before our technology actually does."

Jose grinned. "Yes, well, I spend most of my time in the vineyard, and I know it very well. I listen, watch, and learn." He paused and looked out across the verdant green vineyard. "For example, with one of our previous winemakers, I told him I thought we were getting a powdery mildew in one block, and we should spray. But he looked at his computer and told me it didn't say so in his logs. Then a few days later, it was crazy with all of the people we brought in to spray when he realized it was an issue."

Michael laughed, and his blue eyes sparkled in the bright sunlight. "Yes, Jose is much more accurate than any of our vineyard technology."

Jose nodded solemnly. "Yes, I see and watch. I am the guardian of the vineyard."

Signature Wine: Hanzell Chardonnay

Though Hanzell makes both Pinot Noir and Chardonnay wines, it is their unique Chardonnay, made in an old-world style with crisp acidity, mineral notes, and a very long finish that most sommeliers will name as an excellent choice to pair with food. Hanzell Chardonnay is also known for its ability to age well and is, of course, created from the oldest

Chardonnay vines in North America, along with blended lots from other sections of the vineyard. Therefore, the wine most representative of Hanzell Vineyards is their Chardonnay.

After the vineyard tour, Michael invited me to visit the new winemaking facilities up the hill and to the northwest of the tasting room. I was impressed to see how much of it was located outside, with large stainless-steel fermentation tanks lined up under an open rooftop next to the barrel room.

"As you can see," said Michael, "we have an open cellar space. I like this because I think it brings in the spirit of the place and vineyards into the winemaking process. Of course," he laughed, "there are some issues with an open cellar, such as leaves and fruit flies."

We wandered through the row of stainless-steel tanks on a cement walkway until we reached a flight of steel mesh stairs. Michael motioned for me to climb up first.

"The loading dock is up here," he said. "This way we can handle the grapes gently and use gravity flow."

As we mounted the stairs, I asked, "How many employees assist in the winemaking process?"

"We have a small winemaking team here—only six people, and we all get our hands dirty during the winemaking process. Bob Sessions, the owner, always helps sort the grapes. We all pitch in with the punching down, pulling hoses, and pressing off juice."

At the top of the stairs were a large cement platform and a several pieces of equipment. "This is where they bring in the grapes after picking in the vineyard," said Michael. "We keep each vineyard lot separate and set up sorting tables here where we perform a hand sort to ensure that any leaves and other materials are taken out before we put the Chardonnay grapes in tank."

"But don't you crush and press the grapes first before putting them in tank?" I asked, referring to the standard method most wineries use to make Chardonnay.

"We actually crush them with the stems but do not press off the juice.

Then I put them into a tank for a two-hour cold soak, before pressing off the juice."

"Why do you do that?"

Michael smiled. "I find this adds more intensity and structure to the wine and allows it to linger longer on the palate. My goal here is to create a Chardonnay that will age at least ten years. Next, we let the juice settle in tank before racking off and transferring thirty percent to new French oak and the other seventy percent to tank."

"Do you add yeast?"

"Yes, we usually use Montrachet or 3309 yeast to start fermentation. All lots are fermented separately. After fermentation, the wine in barrel also goes through ML (malolactic fermentation), but the wine in tank does not go through ML."

"So you can keep the crisp acidity and fruit flavors of the tank wine and blend with the oaked wine?"

"Exactly," he smiled. "This way I can keep a naturally high acid in the wine to assist with longer aging but add some complexity and toasty notes with the barrel-fermented wine."

"Do you add ML bacteria to the barrels?"

"Yes we do, because Hanzell was the first winery to capture and propagate ML bacteria, which we patented as ML34. Therefore, we believe in adding our own ML to our wine. After one year of aging, we move the tank wine to neutral oak for six months and the barrel-aged wine to tank. Then after eighteen months we blend, let the flavors marry for one month, and then bottle. Would you like to taste it?"

"You bet."

Back in the tasting room, Michael poured the 2009 Hanzell Vineyard Chardonnay, of which 2,074 cases were produced. As I lifted the glass, I could see the pale yellow liquid with a hint of green at the rim. Swirling it gently, I lifted it to my nose and was treated to an aroma of soft pear and a whiff of vanilla and toast. On the palate, the wine was vibrant with a zesty acidity and notes of lemon and green pear. The finish was quite long, and complex notes of minerality and well-integrated oak lingered on my palate. The wine was very elegant and similar to a fine white Burgundy from France. It did not resemble the stereotypical ripe and tropical California Chardonnays with creamy body and generous oak.

"This wine is a true representation of the 2009 vintage, in that it is elegant with a crisp acidity," said Michael.

I nodded, remembering that 2009 started off with a warm January followed by a stormy, wet spring and then summer temperatures that were not overly hot. This allowed for the fruit to ripen and maintain crisp acidity and fresh fruit flavors, which were reflected in this elegant glass of Chardonnay.

After thanking Michael and Jose for their time, I drove back down the winding road from the tasting room, past the Ambassador, Zellerbach, and Day Blocks spreading out like a green velvet blanket to my right and tried to tap into my feelings about Hanzell Vineyards. Diversity was the word that Michael and Jose used to describe the place, noting how each section was different and provided a unique component for the final blend. This made me recall my visits to Burgundy where the winemakers there also described how a small section of vines in one part of their vineyard could perform and taste so different than another section, based on soil composition, sunlight, and weather patterns that year.

Yes, I thought, this was exactly what was occurring here at Hanzell. This special jewel box of a vineyard, sparkling across the foothills of the

Mayacamas, was mirroring the nuances of the soil, the reflection of the sun, and the temperament of nature that year. And in doing so, each year this vineyard was able to produce diverse and complex wines with exquisite elegance and complexity that spoke of a sacred place.

Fall

"Harvest & Autumn Leaves"

Diamond Creek Vineyards

Beaulieu (BV#1) Vineyard

Chapter Eleven

Diamond Creek Vineyards
Diamond Mountain AVA, Napa Valley

Just south of the charming town of Calistoga and about a half mile up the tiny and winding Diamond Mountain Road is the beautifully engraved gate of Diamond Creek Vineyards. Covered with gold metal leaves, the tall, black iron gate is arched on the top and opens in the middle. It is flanked on both sides by tall pillars made of river stones and long curving rock walls.

I arrived in front of the gate on a late September morning when the air was crisp with the first scents of autumn, and the leaves on some of the nearby trees were starting to show hints of gold. There was no sign announcing that this was the entrance to the famous Diamond Creek Vineyards, but I knew from the address on the metal plaque affixed to the rock pillar on the left side of the gate that I was at the right location.

Picking up my cell phone, I was starting to punch in the number for the winery when a tiny woman emerged from a wooden house set back in the fir trees to the left of the gate. She was wearing a tan denim jacket over black pants with moccasin-style flats. Her curly, red-blond hair caught the sun, and as she hurried toward me, I realized she was Boots Brounstein, owner of Diamond Creek Vineyards. In her seventies, she still looked very young and energetic with a quick step and a bright smile.

"Welcome," she called out in a vibrant voice with musical undertones. "Are you here for the ten o'clock appointment?"

I nodded yes, and after brief introductions, she smiled broadly and gestured to the gate. "You know it will open if you drive a little closer and whisper 'Open Sesame.' That's what I always tell my grandkids." Her blue eyes sparkled mischievously, and she almost seemed like a fairytale creature with her delicate features and wise words on how to enter an enchanted kingdom. "Go ahead up to the winery," she continued. "I'll meet you there in a few minutes.

So I inched my car closer to the gate and murmured "Open Sesame," and the beautiful black-and-gold gate slowly opened inward, and I traveled uphill on a small paved road surrounded on both sides by tall green fir trees, brambly oaks, and red-barked manzanitas. The foliage was so dense I couldn't help but wonder how Al Brounstein, the founder of Diamond Creek Winery and Boot's deceased husband, could have had the vision and energy to clear these hills of the dense growth and plant twenty-two acres of what would become one of the most legendary vineyards in California.

Arriving at the top of the hill, I saw a green swath of vineyard with multiple ripe purple clusters hanging from each vine. To the left was a large peach-colored building, which I recognized from photos to be the winery. When I entered, Phil Ross, the general manager and Boots's son, approached to greet me. Dressed in a black shirt, tan pants, and a pair of tennis shoes, he looked equally ready to walk through the vineyards or greet visitors from around the world. He had a warm and engaging air, with light-blue eyes, silver hair, and a ready smile.

"Welcome," he said, "would you like to see the view?" He ushered me toward a large bank of windows at the far end of the room, and as I walked forward, I realized the building was perched on the edge of the hill. As I reached the end of the room and looked out the window, I gasped.

Nowhere in Napa Valley was there such a magnificent sight. It reminded me of my first glimpse of the Grand Canyon, when after driving for miles through flat land, I finally reached the edge of the canyon and looked down to see cascading rose-colored cliffs and the winding green Colorado River far below. Here, instead, after driving through winding tree-covered hills, I was now overlooking a vast valley of vineyards with a small creek winding below interspersed with a lagoon, waterfalls, and gardens filled with multihued flowers.

Phil laughed. "We often get that reaction when visitors see this view for the first time," he said. "In fact, Al Brounstein had a similar reaction the first time he stood here on the hill and looked down at the valley— only there weren't any vines then."

History of Diamond Creek Vineyards

It all started when Al Brounstein, owner of a district drugstore, took a wine-tasting class at UCLA in the early 1960s and became enamored with the idea of planting his own vineyard and making fine wine. Fortunately he was asked to join the Hollywood Wine and Food Society where he met Dick Foster, one of the owners of Ridge Vineyard. Dick encouraged Al in his dreams to plant a vineyard and invited him to help harvest at Ridge Winery in Santa Cruz several times. It was while working part time at Ridge that Al became convinced that he wanted to plant Cabernet Sauvignon on a hillside.

After his first wife sadly died, Al met and married Boots Brounstein. Because Al was an avid pilot, the two of them, along with twelve-year-old son, Phil, would fly up to Santa Cruz and Napa Valley in Al's small plane looking for a location to plant the vineyard. During this time, Al consulted with many wine experts, including André Tchelistcheff and Louis Martini, who both encouraged him to look for land in the Napa Valley.

Eventually his real estate agent found a hillside property of eighty acres just south of Calistoga. Al had doubts about buying it at first because there were very few vineyards in the area, but after riding in the back of a pickup truck and being scratched by the thick manzanita tree branches that crowded the narrow dirt road, he saw a beautiful valley with a small stream running through it and knew it was the place for his vineyard. Because the stream was filled with small pieces of quartz crystals that glittered like diamonds, Al decided to name his vineyard and winery Diamond Creek.

View of Vineyards from Winery

He purchased the land in 1968 and used an old tractor to clear the brush and trees on the top of what is now the Volcanic Hill Vineyard. According to Boots, "Al was a visionary, but he also had a halo around his head. Do you know he bought the land and wasn't even sure if it had enough water for a vineyard? Yes, he knew a creek ran through the property, but it was only later that we dug a well and fortunately found water. He also had no idea that there were different soil types here until he cleared the land. Where else in the world do you find three distinct soils coming together around a creek—and only sixty feet apart?"

Once Al cleared the land and found the three different soil types, he named each part of Diamond Creek Vineyard after the particular geology in that section. Therefore, the top of the hill with the fluffy, gray, volcanic ash soil became Volcanic Hill (eight acres) and the iron-rich red soil with the small rocks became Red Rock Terrance (seven acres), whereas the gravelly soil in the meadow was christened Gravelly Meadows (five acres). Several years later he created a small lake and planted a three-quarter-acre block that is known as Lake Vineyard, as well as one acre of Petit Verdot. Therefore, today Diamond Creek Vineyard is composed of almost twenty-two acres of vines.

Always the visionary, instead of buying vines from California nurseries, Al used his connections in France to smuggle in Cabernet Sauvignon and Merlot budwood from Bordeaux.

"Al received the budwood from two of the famous First Growth wineries," explained Boots, "but we made a promise to never divulge their names." Therefore, Diamond Creek is planted on vines that originally came from Latour, Lafite, Margaux, Mouton-Rothschild, or Haut-Brion. The vines were shipped to Mexico, and Al flew down in his plane where he smuggled them over the border. He then hired a crew to graft the budwood to St. George rootstock.

Next, Al and Boots hired Jerry Luper as their consulting winemaker, and Jerry agreed to ferment and bottle the wine from each vineyard separately. However, in the beginning Al had problems trying to convince wine shops to carry all three labels.

"They told him they didn't have room on their shelves for all three vineyards and that he should blend them together," recalled Boots. "However, Al told them that as soon as DRC blended their La Tache and Romanée Conti Vineyards together, then he would consider blending his." She smiled brightly with a humorous glint in her blue eyes. "That stopped them, and then they agreed to stock all three vineyards."

That was in the early 1970s when the wine was retailing for $7.50 per bottle. Today Diamond Creek Winery is one of the most widely collected and coveted California Cabernet Sauvignon with retail prices ranging from $200 to $500 per bottle. Diamond Creek has received countless stellar reviews from wine critics over the years and is considered to have excellent aging capability.

In the beginning, the Brounsteins made their wine outdoors in open-top tanks and stored it in barrels at other winery facilities until they built their own winery in 1980. When Jerry Luper retired in 1991, the Brounsteins hired Phil Steinschriber as winemaker and vineyard manager.

Touring Diamond Creek Vineyards

"So you found your way here," a musical female voice said from behind me. I turned to see Boots standing nearby with two cups of coffee. "I thought we would start in the office and have some coffee. Is that all right with you?"

"Of course." I followed her into a large, comfortable room with a desk, computer, and chair on one side and a small living room complete with leather sofa, two chairs, coffee table, and a fireplace on the other side. Over coffee, Boots and Phil described some of the early years when they were establishing the winery.

"It was challenging in the beginning," said Boots. "We used to fly up here from southern California on the weekends with Al piloting the plane, and check into the Calistoga Inn. At that time there were no traffic lights in Calistoga or St. Helena. It was quite rural."

"What did you do on those weekends?" I asked.

"Al would spend most of his time in the vineyards with Sergio, who we hired as a full-time vineyard worker. He is still with us today," Boots said proudly with a soft smile on her face.

"I remember having to move the irrigation pipes around in the vineyard on those weekends," said Phil with a grin. "I was only sixteen, and it was a lot of work. We didn't have drip irrigation back then."

"Yes," said Boots with excitement in her voice, "and in the beginning when the vines were young, we had to bring in a water truck and water them by hand. Then Al had trouble with the rabbits eating the leaves, so he stopped by the dairy in Sonoma and bought a bunch of milk cartons to protect the vines. I still remember when Fred McCrae from Stony Hill came to see the vineyard. He said, 'It looks like a cemetery of milk cartons.' Al was rather hurt by this because he was so proud of the vineyard."

"What else did you do during those early years?"

Al and Boots circa 1995

"Al loved gardening," said Boots, "and he started to build the lagoon, pond, waterfalls, and rock walls. He always had an artist's eye and hated straight lines. When Sergio was putting in the rock walls, Al would often make him take out the rocks—even though they were already set in cement—so that the walls would curve."

"Tell me about the waterfalls," I requested. Diamond Creek Vineyard is famous for the thirteen waterfalls that grace the property.

"Al was fascinated by waterfalls," Boots smiled. "He started building water features on the property right away. He was always improving things. I still remember the day he asked me to take a walk with him to the back of our property so he could show me something. We pushed our way through a wilderness of trees and shrubs, and then he told me he wanted to create a lake there. I said, 'Al you've lost it,' but he was such a visionary, he could clearly see how that tangled spot could become a beautiful lake."

"We've built some hiking trails around the lake," said Phil. "We invite our visitors to walk through the gardens and vineyards, past the waterfall park, and picnic or relax near the lake."

"I remember in the early days finding many pieces of quartz in the creek," said Boots. "We also found arrowheads on the property, which were made by the Wappo Indians from black obsidian." She handed me a framed box with a glass lid. Inside, nestled on red velvet, were an artistically arranged collection of arrowheads and sparkling white quartz stones.

"Very beautiful," I said. "Fascinating to think of the Wappo Indians who used to live in Napa Valley. Tell me about your early experiences with making wine here."

Boots laughed and jumped to her feet. She walked to the window and pointed down into the vineyards. "See where the creek runs to the right of the lagoon. Well, just beyond the trees there, we set up an old redwood tank. We used to pick the grapes and crush in the vineyard."

"At one point they stored the barrels in some old caves further up the mountain," explained Phil, "but eventually, when they built the winery near the gate, the wine was stored in the garage."

"Yes, and one year we didn't have enough room in the garage to bottle it," said Boots. "So the TTB came out and bonded our driveway. That way we could bottle the wine in the driveway and still be legal." She laughed, and her blue eyes sparkled. "I think we are the only property in Napa Valley that has a bonded driveway."

"This winery location was finally built in the 1980s," said Phil. "Later we added these offices on the top floor. Are you ready to take a ride around the property?"

"Yes!"

As Phil led me outside, I could look down to the winery loading dock below. It was filled with many boxes of newly picked grapes that were being loaded into the destemmer for processing. Workers swirled around the equipment, rushing to get the grapes processed in the cool of the morning.

"As you can see," he said, "we started picking this morning and are now crushing the grapes. These were from a warmer portion of Red Rock Terrance. In the end we will do multiple passes through the vineyard because the various microclimates ripen at different times."

He ushered me into a six-passenger golf cart painted in dark green with beige trim, and then we headed down the hill and into a parklike area. It was filled with tall, leafy green trees, a small brook with a delicate waterfall tumbling through the center of it, and several picnic tables. The whole grove exuded an air of calm tranquility.

"This is one of Al's first waterfall parks," said Phil. "We invite friends and employees to come here and relax."

"Beautiful," I murmured, as a feeling of awe and peace enveloped me. In all my travels to vineyards of the world, I had never encountered a location where the vines were integrated so beautifully into a parklike setting.

We continued driving for several yards, and then Phil stopped the cart near Gravelly Meadows Vineyard. Here the soil was a light gray-brown color and strewn with small pebbles and rocks. The vines were widely spaced with thick gnarled trunks and dripping with purple clusters of grapes. I jumped out of the cart and walked into the vineyard, reaching down to scoop up a hand of the gravel. Very similar to Bordeaux's gravelly soil, I thought.

"Help yourself to a berry," said Phil, as he sampled one from a nearby vine.

I approached an old vine that was partially covered with green moss, highlighting its age. The leaves were a glorious mixture of greens, with some starting to change to yellow, red, and orange. The loose clusters of Cabernet Sauvignon were a brilliant purple with navy-blue hues, and seemed to arrange themselves artistically like ornaments on a carefully hung Christmas tree.

Old Vine in Gravelly Meadows Vineyard

Reaching out to one of the hanging clusters, I gently picked a berry and placed it in my mouth. The flavor was sweet and jammy on my tongue, and the texture of the skin was velvety. The acid was strong yet balanced. I spit out the seeds and saw they were half brown and half green.

"They just need a little more time," said Phil with a smile on his face. "Our winemaker, the other Phil, and I sample these grapes almost every day during this time of year to determine when to harvest each block."

"I see these are cane-pruned vines rather than cordon," I said, touching a long slim cane that stretched upward, attached to horizontal wires. It had two perfect purple clusters and was surrounded by other long canes that were also perfectly balanced with two clusters.

"Yes, we are using both systems here at the estate. Al put many of the older vines on cane, especially in the cooler areas such as Gravelly Meadow and Lake Vineyard where we find cane works better. In a few minutes, you will be able to see that we are using VSP (vertical shoot positioning) with cordon and Geneva Double Curtain trellising in the warmer parts of Volcanic Hill and Red Rock Terrace."

"How old are these vines?" I asked, very impressed with the width of the shaggy trunks.

"Most of them are over forty years old and were planted when the vineyard first started. We don't pull out old vines unless they no longer produce. Then we use our own budwood to graft over a new vine and plant it among the old." Phil stopped and pointed down the row. "See, there is a new vine there."

I glanced over, and sure enough, there was a tall, slender vine growing valiantly amid the ancient beauties that dwarfed both sides of it. However, since the original twelve foot by eight foot spacing was kept, the new vine had plenty of sunlight and room to grow.

Back in the cart, Phil drove to the edge of the lagoon and parked near a small bridge. "Let's walk over the bridge to see Volcanic Hill," he suggested.

Starting across the bridge, I realized there was a small island in the middle of the lagoon. It was covered with pink-and-red rose bushes and fragrant purple wisteria blossoms in full bloom, even in the autumn season. As I walked, I could smell the sweet aromas of flowers filling the air and saw white butterflies flitting among the blooms. A tall post with three signs pointed the direction to the three vineyards, each with the information that the designated vineyard was only sixty feet away. It was fascinating to realize that this lagoon and island stood at the center of where the three distinctive soil types met.

I walked across the second half of the bridge and was at the edge of Volcanic Hill Vineyard. The difference in the soil was immediately visible in the large mounds of fluffy white-gray soil surrounding the foot of the old vines, which stood tall and proud, stretching their cordon limbs four feet in both directions. Dark purple-black grapes peaked from beneath green leaves. I couldn't stop myself from kicking a mound of the fluffy soil, and it responded by producing a satisfying cloud of gray dust.

Signs Pointing to Three Vineyards

"As you can see," said Phil. "Not only is the soil different, but it is warmer here, and it gets even hotter as you climb farther up this eight-acre block."

We stepped back into the cart, and the next stop was the middle of Red Rock Terrace. It was difficult to believe we had only traveled a few yards, because the soil had changed from a fluffy white-gray color to a warm red-orange hue. The vines marched up the hill on terraces, which were strewn with hand-sized red rocks.

"This is Red Rock Terrace," said Phil, "and we are using the Geneva Double Curtain (GDC) trellis system here to open up the canopy. The vines in the red soil are more prolific, and they produce more leaves, so we needed a method to get the sun to the grapes."

"Interesting," I said. "So the soil is more prolific here and has some iron content." I bent down to scoop up a handful of the rich red earth. It felt warm in my hands, and I was once again amazed at how different and yet how close together geographically these three distinct soil types were.

Even more impressive was how they impacted the taste of the wine. This red soil of Red Rock Terrace was said to create softer, more approachable tannins and brighter red fruit flavors in its Cabernet Sauvignon and Merlot grapes, whereas the pebbly soil of Gravelly Meadows was credited with producing wines with fine-grained tannins but with more earthy notes and darker fruit. The fluffy soil of Volcanic Hill was said to produce the biggest and most austere wines of all three, with massive tannins, complex herbal notes, and black fruit, and it usually required more bottle aging before consumption.

According to *Al Brounstein's Oral History*, recorded and published by the Bancroft Library at UC Berkeley several years before his death, Al described the character of the three vineyards as follows:

> *Although I won't tell you where I got the vines, even though I'm going to tell you three names out of the five (Bordeaux first growths), of which two may or may not be included...I'm going to say just for characteristics of the wine...the Red Rock Terrace...it's relatively soft and easy to drink, it has a lower tannin feel...we call that our Château Margaux. But our Gravelly Meadow is full of gravel, very much like the*

soil of Graves, south of Bordeaux, and that is renowned for its Château Haut-Brion. But our big wine is the Château Latour…Volcanic Hill. That is a biggie, it has long life potential…and it's magnificent. (p. 26).

After leaving Red Rock Terrace, Phil drove the cart awhile longer, and I realized we were passing the lagoon again and the north side of Gravelly Meadow.

Vines in Red Rock Terrace

"I'm taking you to see Lake Vineyard," he explained, "but first I thought we would stop at Waterfall Park, and you can walk to the lake from there. I'll meet you on the other side. Just follow the path."

The song "Follow the Yellow Brick Road," ran through my head as I started up the dirt path that wound through a grove of trees and past more picnic tables. Then a series of steps appeared as the path veered sharply up a hill to the left. The whole experience at Diamond Creek seemed to be taking on a glow of the enchanted, as reminders of fairytales and classical stories continued to surface.

At the top of the hill, I was surprised to see a cliff wall of dark gray stone, interspersed with a series of beautiful waterfalls streaming like lacy white veils down the side of the rock. The path continued in this direction, and I lost count of the number of waterfalls I passed, of all different sizes and producing soft trickling or roaring water sounds as they tumbled down the cliffj.

Eventually the trail rounded a corner, and I could see the blue shimmer of water in the distance. The shimmer became broader, and the navy-blue-and-green body of the lake came into view. It was a small lake, surrounded on three sides by tall fir trees. The trail veered to the left and passed through a grove of red-barked manzanitas glowing in the sunlight, and then I was on the main road again and could see Phil waiting in the golf cart near another small vineyard on the opposite side of the lake.

"How was your walk?" asked Phil.

"Great! Exhilarating. The waterfalls are marvelous!"

"Glad you liked them." He gestured to the vines behind him. "As I'm sure you've guessed, this is Lake Vineyard. It was planted in 1972 and is only three-quarters of an acre. It has our coolest microclimate with the ocean breezes coming over the mountains each afternoon and cooling down this area. The vines are cane-pruned on VSP trellis here, and we are using eight foot by eight foot spacing here because Al thought it was more pleasing aesthetically."

"So this is the famous vineyard that only produces wine every few years?" I asked.

Phil nodded. "Yes, it is so cool that we often can't harvest the grapes because they don't get ripe enough. We have only made wine from this vineyard fourteen times since 1978, but when it does ripen, it creates phenomenal long-lived wines with deep, rich texture."

I walked into the rows to sample a few grapes and found they still tasted slightly herbal with a hint of green pepper. They were not yet ready to be harvested. The soil in Lake Vineyard appeared to be similar to Gravelly Meadow, with gray-brown dirt and small gray rocks.

Phil pointed across a small creek running next to the vineyard that fed into the lake. "Just across the creek, we've planted some Petit Verdot," he said, "and we just built a new hiking trail on the opposite side of the lake. Are you up for another short walk?"

Lake with Lake Vineyard on Far Side

"Sure," I said, excited to be outside on such a beautiful day and hiking in the woods of Diamond Mountain. As we walked closer to the lake, I could see lounge chairs strategically positioned at quiet intervals around the lake, as well as a small canoe on the right bank. A patch of green, velvety grass next to the water with two chairs underneath the shady branches of a delicately leafed tree looked especially inviting, and I could image relaxing there on a sunny afternoon with a good book.

Phil started up a rather steep dirt trail, but there were handrails to hang on to as we climbed into the forest above the lake. The trees rose tall on all sides of us with thick underbrush, but we could see glimpses of the sparkling lake water shining through the foliage. As we hiked along the well-marked trail, I was impressed once again with the visionary spirit of Al Brounstein who created this lake and the legendary vineyards out of such dense brush and trees in what was, at the time, just a mountain wilderness.

After about ten minutes, we arrived at the far end of the lake and found the original trail I had taken a short time before, which ran along the base of the waterfall cliff. We crossed a small dike at the end of the lake and then walked back through the red manzanitas where we could see the golf cart waiting for us next to Lake Vineyard.

Vineyard Specifics for Diamond Creek

As we continued back toward the winery, Phil provided more details about the vineyards. "In terms of *varietals* on the estate," he said, "around eighty-eight percent of it is Cabernet Sauvignon, with the remainder divided between Merlot, Cabernet Franc, and Petit Verdot. We were the first in Napa Valley to plant Petit Verdot in the 1990s. Al felt it not only added excellent color to the blend but also provided structure and some dark fruit and violet notes." Phil paused and glanced out over the vineyards. "You may recall that Boots told you how Al planted these as a field blend?"

I nodded, remembering her colorful account of Al's visits to the vineyards of France.

"Al spoke fluent French," she announced proudly, "so when he was in France, he talked to the French vineyard workers, and they told him the vine varietals were mixed in the field. Therefore, when Al planted these vineyards, he mixed his varietals, but he knew his vines so well he could name each one."

The *soil* composition of Diamond Creek Vineyards is known as Napa Volcanics, which according to the authors of *The Winemaker's Dance*, can include "tuffs, lava flows, mudflows, pyroclastic flows, and stream deposits (p. 38)." This helps to explain the wide variance of Diamond

Creek Vineyards with its three distinct soils of fluffy gray dust known as the Forward Series; the iron-rich red soil from the Boomer Series, and the gravelly well-draining soil of the Felta Series.

Vineyard Specifics for Diamond Creek

Total Vineyard Acres	22 (Volcanic Hill: 8; Red Rock Terrace: 7; Gravelly Meadow: 5; Lake: 3/4 plus 1 acre of Petit Verdot)
Varietals	Cabernet Sauvignon (88 percent), Petit Verdot, Cabernet Franc, and Merlot (mixed field blend)
Soil	Napa Volcanics: Red Rock Terrace from Boomer Series, Volcanic Hill from Forward Series, and Gravelly Meadows from Felta Series
Elevation	800 to 1,000 feet above sea level
Average Temperature	Summer temperatures average mid 90°F for a high and mid 50°F at night. Rainfall is 40–55 inches per year.
Rootstocks	St. George
Clones	Bordeaux First Growth clones
Sun Exposure	East/west orientation (old method)
Spacing	12 x 8 for older vines, 8 x 8 on Lake, 7 x 4 for newer plantings
Trellis Systems	VSP, Geneva Double Curtain; both cane- and cordon-pruned

Elevation ranges from 800 to 1,000 feet above sea level, and average summer *temperature* is mid 90°F in the day with a night time low of mid 50°F. The average rainfall for the Diamond Creek AVA is 40 to 55 inches per year,

Rootstock is primarily St. George on all of the older vines. St. George was selected, "because I knew that it was a drought-resistant rootstock (p. 44)" reported Al Brounstein in his *Oral History*. *Clones* are derived from the Bordeaux First Growth budwood that Al had shipped from France to Mexico.

In terms of *sun exposure*, the vines "should probably be north by south," explained Phil, "but the old way of planting vines was by the contours of the land. Therefore, most of our older vineyards are on an east/west orientation."

The majority of *spacing* is twelve by eight feet, as it was originally laid out so it could accommodate tractors of that time period, but there are a few newer sections where spacing is seven by four feet. The Lake Vineyard was purposely laid out as eight by eight, because Al felt the wider twelve by eight spacing would show too much dirt and not be as scenic. According to his *Oral History*:

> "If I had twelve-foot rows, it would be easy to get the tractors up and down, but it would look like heck. All you'd see would be all that dirt. If you were on the lake paddling around...it would look awful...So I planted eight by eights...The heck with the soil; I want the beauty (p. 32)."

The *trellis systems* are VSP and Geneva Double Curtain (GDC), with a combination of cane and cordon training. The different systems are used to accommodate the various microclimates within the blocks, with VSP trellis and cane pruning used in cooler areas such as Gravelly Meadows and Lake Vineyards and a combination of VSP cordon pruned and GDC in the warmer areas.

Farming Practices—Sustainable

Diamond Creek Vineyards utilizes sustainable farming methods but is not pursuing certification at this time.

"We try to use organic and natural products in the vineyard whenever possible," explained Phil.

Fertilization and *weed control* are primarily organic with a natural cover crop allowed to grow in the winter and spring. It is then mowed in the summer and disked into the soil. According to Phil, "we cut weeds with a Weed Eater and make compost each year, which we put back on the soil after harvest."

Farming Practices at Diamond Creek

Certifications	Utilizing California Sustainable Farming practices but not certified
Fertilization	Natural cover crops and organic compost
Weed Control	Mowing and Weed Eater
Canopy Management	100 percent by hand: suckering, thinning, pruning, harvest
Disease Control	Sulfur and other organic products to prevent powdery mildew
Pest Control	No major issues, using owl boxes
Irrigation	Dry-farmed in Gravelly Meadows and Lake Vineyards; double-drip irrigation system on newer vines
Technology	Petiole analysis and pressure bombs; frost not a problem
Harvest Measurements	Winemaker and general manager determine by taste and lab readings

Canopy management practices are implemented 100 percent by hand. "We continually go through the vineyard," reported Phil. "We are later than the rest of Napa Valley because it is cooler here. Therefore, we usually sucker in April during budbreak. We thin to two clusters per shoot, and at verasion in July, we may trim shoulders and wings."

Disease control includes the use of sulfur and other organic products to prevent powdery mildew. *Pests*, such as deer, are not an issue because the property is deer-fenced. According to Boots, "there are still plenty of turkeys, rabbits, peacocks, gophers, birds, and foxes in the vineyard, but they are not much of a problem in terms of eating the grapes or vines. We have put owl boxes through the vineyard, but we don't have any owls in them yet."

For *irrigation*, dry-farming is used in Lake and Gravelly Meadows Vineyards, which are cooler, but drip irrigation is employed in the warmer regions of Volcanic Hill and Red Rock Terrace. Phil pointed out the unique double irrigation system of two sets of black hoses running along the bottom trellis wire closest to the vines.

"We installed this system because we replant so many new vines among the old. We need to water the new vines more, so we program these double irrigation lines so that each vine receives the amount of water it needs."

I was very impressed with this system, because not only does it help the young vines thrive and preserve the old, but it is also a clever water-saving mechanism.

Other *technology* used in the vineyard includes using pressure bombs to test water needs, as well as petiole analysis each year to analyze vine nutrition. Frost fans and heaters are not needed because, according to Phil, "frost is not a problem here. We are in a bowl, and it is warmer. We also get a nice afternoon breeze."

Harvest measurements are determined jointly between the two Phils. "We both walk the vines," explained Phil, "tasting the grapes and collecting enough to do a sample in the lab. We generally make our decision on when to pick based on taste but will also consult the lab numbers. Harvest can last over a month here because the vineyards ripen at different times, and we may make four to five passes in each vineyard. Volcanic Hill is first because it is warmer, then Red Rock Terrace and Gravelly Meadows, and Lake is last. Harvest usually starts now—at the end of September—and we generally don't finish until the end of October."

Phil paused and glanced out over Gravelly Meadows Vineyard. "We flower and harvest later than other Napa vineyards," he said. "Our vines are still sleeping when the others are coming out. We find this to be a positive and never have to worry about the Brix being too high and the fruit not ripe. This is a blessed spot for Cab."

Economics of the Vineyard

The *average yield* of Diamond Creek Vineyards is a very low 1 to 2.5 tons per acre. With only 22 acres, this equates to a production range of 22 to 55 tons per year, or a *total average* of only 39 tons per year.

Economic Viability of Vineyard

Average Yield	1–2.5 tons per acre
Total Average Tons Per Year	39
Costs Per Acre	Not available
Revenues	Bottle prices range from $200–$500
Economic Health	Very good

Though farming costs per acre are not available, Diamond Creek has a loyal vineyard workforce that has been tending the land for many years and enjoys a reputation for treating their employees well. Currently they have twelve full-time employees, of whom five work in the vineyard. Of these five vineyard employees, all have been there a minimum of twenty years, with Sergio, the vineyard supervisor, having worked at Diamond Creek for over forty years. He was the first employee to be hired by Al

Brounstein.

In terms of *revenues*, the winery produces around two thousand cases per year with an average retail price of $200 per bottle for the three major vineyards. When Lake Vineyard does produce one of its rare vintages, the average price per bottle is around $500. Diamond Creek does not sell any of its grapes, but like the great Bordeaux châteaux, they may declassify a portion of their wine that they do not believe is up to par for a particular vintage.

As an allocated winery, Diamond Creek has the luxury of deciding to whom they are going to sell their wine. According to Phil, "We establish relationships to decide who gets the wine." Even though their production is quite small, they still manage to sell their wine in forty-five states and in many countries around the world. Phil rates the *economic health* of the vineyards and winery as "very good."

Soul of the Vineyard—"Ohhhh" and "Serenity"

Phil parked the golf cart in front of the winery and ushered me into a room with a long wooden table and intricately carved wooden chairs. Boots was already seated at the table, and she welcomed us back, encouraging us to sit down, have some water, and help ourselves to the tray of cheese and crackers that was artistically arranged on the table, along with plates, napkins, and three wine glasses at each setting.

In the middle of the table, three half bottles of the 2009 Diamond Creek Cabernet Sauvignon vintage took center stage with the names of the vineyards proudly announced on each bottle: Red Rock Terrace, Gravelly Meadows, and Volcanic Hill.

"How did you like the vineyards?" asked Boots.

"I've never seen vineyards arranged in such a beautiful setting before," I answered truthfully. "The waterfalls, garden, lake, and vines are all intermingled to create an enchanted place."

Boots smiled. "We like to think so."

"Before we start the tasting, do you mind if I ask each of you to describe the personality of Diamond Creek Vineyard in one word?"

There was silence as they both thought about the question, and then Boots was the first to answer. "You know, we've had hundreds of visitors come here over the years, and they all seem to have the same reaction when they first see the vineyards. They stand by the window, just like you did, and look out across the land, and they say, '*Ohhhh.*' They all seem so surprised and taken back by the view, and then they usually follow the 'Ohhh' with 'how beautiful' or something like that. I have to admit that I feel the same way each time I look at it. There is a sense of awe that takes your breath away."

I nodded in understanding, because I had experienced the same thing.

"You know," Boots continued with a faraway look in her large blue eyes, "When Al was dying of Parkinson's disease and they told him he only had a few days left, he insisted that the ambulance bring him here so he could look at the vineyard one last time. They rolled his bed up to the window there so he could see the vines, and he said, 'Aren't they beautiful?' Those were some of his last words. He had a deep emotional attachment to the vineyard. I think that was why he never wanted to pull out old vines, and if they died, he would replace them with the same budwood."

"Yes," agreed Phil. "Al really loved the vineyard, and even though he had Parkinson's, he didn't see himself as a victim. He didn't let it slow him down, and he even used to make jokes about it to help put people at ease. For example, he'd tell people he had to give up flying his airplane because he had so many rough landings, or that he could only do fast numbers when he was conducting an orchestra."

We all laughed, and then I turned to Phil and asked him if he had his one word yet.

"Mine would have to be serenity," he said. "There is a strong sense of the serene when you are walking around the property. I especially love it in the fall when the leaves in the vineyard turn so many colors—it is just as beautiful as autumn on the East Coast."

Phil Ross, General Manager

Next we moved onto the question of the biggest challenges and joys of working with the vineyard.

"We've had so many challenges over the years," said Boots, "especially in the beginning with money when the banks didn't think we would make it."

"The beginning was tough," agreed Phil, "but another issue is that the wine business is not like producing widgets where you have more control. Here, nature plays a huge role, and you have to be patient, but I also find joy in farming and the changing of the seasons."

This last aspect was also echoed in Phil's perception of what the vineyard has taught him over the years. "I have learned," said Phil, "that you have to take what Mother Nature gives you and treat it with respect. All three vineyards are so very different. You need to be patient and allow them to complete the process in their own time."

Boot's response to what the vineyard had taught her was so poetic that Phil and I were both struck silent with the beauty of her words and the dreamy note in her voice as she spoke. First, she leaned back in her chair, closed her eyes, and smiled. Then she opened her eyes and gazed intently at us.

"Patience," she said softly. "Through the years, I have experienced the vineyard and this place through the seasons. In the winter, the vines are sleeping and getting strong, and in the spring there is no greater thrill than budbreak, when the valley awakens with bloom in the vineyards. Then in summer there is the warm sun and color of the grapes changing, and then we reach harvest and autumn where the leaves change to so many beautiful colors." She looked up and smiled at both of us. "So I've learned patience and know that the cycle will continue."

Signature Wine: Diamond Creek Cabernet Sauvignon

The opportunity to taste all three vineyards side by side was a special treat for me, because on previous occasions I had only tasted Diamond Creek wines in one-off situations. Over the years Diamond Creek wines have scored quite high with wine critics, with some vintages receiving rave reviews. For example, the 1978 Diamond Creek Volcanic Hill is considered a legendary wine that many experts compare to the best First Growths of Bordeaux.

The 2009 vintage in Napa Valley is known as a cooler one but with heat spikes during August and then a huge rainstorm on October 13 that caused problems for some wineries. This did not appear to be the case for Diamond Creek, as all three wines were exquisite, though with very different personalities.

Phil encouraged me to start tasting the wine in whatever order I wanted, so I decided to follow the standard wisdom of tasting from the lightest to the heaviest tannin structure. Therefore, I started with the 2009 Red Rock Terrace. As I lifted the glass to my nose, warm raspberry and spice notes immediately assaulted me. On the palate the tannins were velvety, and the fruit deepened to show black plum, anise, and a touch of pepper. Like Margaux in a warm vintage, I thought, but ultimately a classic Napa Valley Cab blend. The oak was well integrated and the finish long and elegant.

Next was the 2009 Gravelly Meadow, which was darker in color with some black hues among the ruby red. The nose was much earthier with dark fruit and spice. This carried through on the palate with complex blackberry, tobacco, and forest-floor notes. The balance was excellent with a refreshing acidity and fine-grained tannins. More similar to a Pauillac, I thought.

I picked up the glass of 2009 Volcanic Hill with a strong sense of anticipation. As a fan of massive tannins and earthy flavored wines, I was looking forward very much to tasting the vineyard that Al Brounstein referred to as "the biggie." It did not disappoint. In terms of color, it was the darkest of the three, and the nose was a brooding combination of black cassis, mushroom, and pencil lead. On the palate, the tannins were much larger with excellent structure and grip, and a variety of complex earthy notes, cedar, and spice swirled around my taste buds. The finish was very long, inviting me back with its complex character for another sip so I could continue to discover new nuances of flavor. This was my favorite of the three, I thought, and it reminded me of the massively structured wines of St. Estèphe in Bordeaux.

After my tasting with Boots and Phil Ross, I caught up with Head Winemaker Phil Steinschriber. He has been the winemaker at Diamond Creek for over twenty years. A graduate of the winemaking program at CSU Fresno where he received his Masters in Agricultural Chemistry, Phil's first winemaking job was in the Golan Heights of Israel producing kosher wines. It was here that Al and Boots first met him, and when Phil returned to the United States several years later, Al offered him the job as winemaker because Jerry Luper was retiring.

Phil started in 1991, working alongside Jerry awhile before taking on the role of head winemaker. Over the years, Phil has managed to receive high scores from wine critics on his winemaking prowess, including many ratings of 90 and above. According to Al in his *Oral History:*

"(Phil) is a very qualified man...He worked for other wineries, and he has a very sharp talent. We trust his views to a great degree (p 41)."

Phil Steinschriber has a look of the Old World about him, with thick hair that is cut in a longer European style, complimenting his generous mustache and beard. He generally wears a baseball cap and has a serious and calm demeanor.

"All the vineyards are distinctively different and offer the flavors of the soils the vines are in and the exposures they experience," he reported.

I was fascinated to listen to his viewpoint, as he has made wine from these vines for over twenty years. "Red Rock Terrace," he said, "is a north-facing, terraced vineyard with iron-rich Aiken soil. The wine resulting from these vines is usually very black cherry/berry in character and very dark and dense. Gravelly Meadow Vineyard, on the other hand, faces slightly east in gravelly loam soils that are alluvial and brought down by erosion from the mountains. I believe the wines resulting from these vineyards show cassis/black currant characters."

Thinking back to my experience of tasting these wines, I had to agree with Phil. The taste profile he described matched my experience very closely.

"Volcanic Hill," he continued, "is a white/beige volcanic ash deposited by the Mt. Konocti volcanic explosion. It is a south-facing slope exposing the vines with its direct aspect to the sun. The wine is usually very dark and brooding with a mineral characteristic complementing the cassis fruit. The wine marries very well with one hundred percent French oak and is considered Frenchlike in character. The wines from this vineyard age for a long period of time."

Phil went on to explain that winemaking methods are designed to match the mood of the vineyard and vintage. Diamond Creek utilizes a rigorous triple-sorting system, similar to those found at the First Growth estates in Bordeaux. The first sorting takes place in the vineyard where workers are trained in how to select the highest-quality fruit. Indeed, they perform several passes through the vineyard, carefully picking only clusters that are at the ideal peak of flavor. The second sorting occurs at the crush pad where two to four workers inspect clusters and reject any that may not be in good condition, as well as throwing out leaves. After destemming and a gentle crush, the berries are sorted a third time by eight workers who reject any green jacks (small green berries that are not ripe), as well as any other berries that may not be in top condition.

Fermentation takes places in two-ton, custom-made, stainless-steel tanks, where depending on the vintage, a cold soak may or may not take place.

"Since this is mountain fruit," Phil said, "the tannins can be huge. Therefore, some years we don't do a cold soak, because we don't want to extract more tannins."

Cultivated yeast is used to start the ferment, which averages around 85°F with manual punch downs performed several times per day. Primary fermentation generally concludes within ten to fourteen days.

"We generally don't do extended maceration, because our fruit is so concentrated," Phil stated.

The must is gently pressed with a hand-operated basket press. Free-run and pressed lots, as well as the different vineyard blocks, are all aged separately in 225-liter French oak barrels with a medium-high toast for twenty-two months. The blending schedule is "dependent on the vintage," Phil explained again, but the wine has time to marry in both barrel and bottle before release.

* * *

As I drove away down the winding road towards the gold-tipped gates of Diamond Creek, I had to pinch myself as if to awaken from a dream. My experiences at this vineyard with its gardens, waterfalls, and butterflies flitting over the flowers on the tiny island in the center of the three distinctive blocks, were enchanting. I thought of Phil and his description of the vineyard as filled with serenity, and I had to admit that I had felt some of that as well.

At the same time, I thought of Boots and her story of Al Brounstein insisting that they bring him from his hospital bed to view his beloved Diamond Creek vineyard one more time before his passing. The view from the second floor of the winery, looking out over the magnificence of the fluffy soil of Volcanic Hill rolling down to meet the rich red-orange dirt of Red Rock Terrace and the corner of Gravely Meadows, must have filled his heart with joy. Just as it does for the few fortunate visitors, like me, who are able to stand at that window and catch their breath in awe at the spellbinding sight.

Chapter Twelve

Beaulieu Vineyard
Rutherford AVA, Napa Valley

November is one of the most alluring months in Napa and Sonoma counties. The vineyards are a vast, undulating carpet of multihued golden leaves accented with an occasional splash of red and orange. The November afternoon that I drove to Beaulieu Vineyard (BV) in the heart of the Rutherford AVA was even more inviting than most, because it was unseasonably warm with a temperature hovering around 78°F. The sky was a brilliant blue, and the sun seemed to etch each golden leaf into perfect detail as I passed block after block of vineyards lining both sides of Highway 29.

When I arrived at Beaulieu Winery visitor's center, the red ivy climbing down the stone wall of the old building was captivating and contrasted beautifully with the tall trees filled with orange-and-yellow leaves that graced the parking lot. Climbing out of my car, I stopped for a few minutes to admire the bronze statue of André Tchelistcheff that welcomed visitors to the Private Reserve tasting room. This was the man who was hired by Georges de Latour, BV's founder, to travel from France to California and create world-class wines that were lauded by presidents.

Entering the tasting room, I asked for Sam Burton, Director of Vineyard Operations, and was directed to walk to the front of the winery offices. Just as I rounded the corner of the building, I saw a slender man of medium height with an athletic build. He beckoned for me to come forward, and as I approached, I could see he was dressed in black jeans, boots, and a blue-and-white-checkered shirt. Sunglasses shielded his eyes from the bright light as he leaned forward to shake my hand vigorously.

"Welcome to BV," he said. "You picked a great day to visit."

"Yes," I agreed, noticing he had a distinct Australian accent and the tanned complexion of someone who has spent many years in the outdoors. "You're from Australia?" I asked.

"You detected an accent?" he asked, grinning.

"Sort of hard not to."

His grin widened. "Yes, well, I grew up in Adelaide and worked in a lot of Australian vineyards before coming here."

"So you enjoy working in Napa Valley?"

"Absolutely. What's not to like about this beautiful place? Are you ready to go see BV1?"

I knew Sam was referring to the first vineyard Georges de Latour planted, which was the oldest of BV's eight estate vineyards. BV1 was also the source of the grapes that went into their flagship wine, the BV Georges de Latour Private Reserve, which was bottled by André Tchelistcheff from the 1936 vintage to honor his employer. Both the wine and vineyard were legendary in global wine circles and praised by wine critics around the world.

Sam ushered me into a large silver Chevrolet truck that was so tall I had to use the running board step to climb inside. Once seated and buckled in, Sam steered the truck into the northbound lane of Highway 29, and then answered some questions about his background.

He had received a degree in Agriculture Science at the University of Adelaide's Roseworthy Campus, which is as well-known internationally as UC Davis University is in California for Viticulture and Enology degrees. Upon graduation, Sam worked in vineyards across Australia as a consultant before going to work for Pacific Wine Partners, which brought him to Napa Valley. When Diageo, owner of a premium group of wine, spirits, and beer companies, offered him a job as director of vineyard operations for all of their wineries, including BV, Sam jumped at the chance.

"So how many acres are you in charge of altogether?" I asked.

"Two thousand acres, but I have many people to help me. In general, I manage to visit BV1 around three times per month and more during harvest."

He turned the truck blinker on and merged into the middle turning lane. Looking out the window, I realized we were less than a mile north of the BV Winery parking lot we had just left a few minutes earlier.

Gated Entrance to Beaulieu Vineyard #1

When there was a break in the traffic, Sam steered the big truck to the left, and we turned into the entrance of BV1. I gasped in surprise at the impressive view. Directly in front of us was a gate designed of green-and-black iron squares flanked by two posts made of bricks and topped with iron lanterns. But it was not the gate that was imposing so much as the long row of sycamore trees that lined each side of a straight, narrow road with vineyards stretching out on both sides. The patchy bark of the trees was a mix of white, beige, and gray, topped by canopies in a golden glory of autumn leaves. The vineyard blocks on the right were showing many green leaves because they were new plantings still receiving irrigation, whereas the vineyard blocks on the left were a blaze of orange, bright-yellow, and crimson-red leaves.

"Wow," I said.

Sam idled the truck in front of the gate and smiled. "Yes, this is a pretty famous view. We actually have it on our winery calendar."

"I can see why."

Just then a loud whistle blew, and the ground started vibrating underneath the truck. I turned my head to the right and saw the Napa Valley Wine Train bearing down on us. It was then that I realized the truck was parked in the middle of the train tracks.

"Guess I should move the truck," Sam chuckled.

"Please!"

He inched the truck forward a few feet off the train track and then rolled down the window to lean out and push a button on the gate controller. The gates slowly opened, and he drove the truck through just as the train roared by on the track behind us.

* * *

History of Beaulieu Vineyard #1 (BV1)

The seventy-eight acres that comprise Beaulieu Vineyard #1 are still owned by the descendants of Georges de Latour, and the family leases the vineyard to Diageo. Diageo, in turn, with the guidance of Sam in the vineyard and Jeffrey Stambor as winemaker, manages the famous vineyard and produces the Georges de Latour wine. In order to track the history of the vineyard, several days before my meeting with Sam, I had been fortunate to meet with Walter H. Sullivan III, the great-grandson of Georges de Latour, who took me on a tour of the property.

When I arrived, Walter had just finished riding one of his prize quarter horses. Dressed in jeans and cowboy boots, he welcomed me with a hearty handshake, and then we walked around the historic property to see the old manor house, gardens, multiple outbuildings, and the horse stables. I was fascinated to realize that BV1 surrounded the whole estate, rather like a protective moat composed of grapevines.

"My great-grandfather Georges," said Walter, "was classically trained in chemistry at the Jesuit College in Paris and not only spoke French and English but was fluent in Greek and Latin."

Georges de Latour in Beaulieu Vineyards

Walter explained that Georges was born in the Bordeaux area in 1856 and sailed to America in 1882 when he was only twenty-six years old. Arriving in New York, he eventually made his way to San Francisco where he worked as a chemist for four years before starting his own company producing cream of tartar from grapes. The business became so successful that he expanded to four locations including San Jose, Healdsburg, Fresno, and Rutherford in the Napa Valley.

In 1897 Georges sold the company to Stauffer Chemical and a year later married Fernande Romer. It was then that he began looking for land to fulfill his dream of planting a vineyard and starting a winery.

"He bought the first fifteen acres of BV1 in 1900 from Charles Thompson," stated Walter, "and purchased the remaining one hundred and fifteen acres in 1903. Interestingly, Charles was a prohibitionist and had owned the land since 1872. It was covered with orchards and wheat, plus there was a Victorian two-story house. When Georges brought my great-grandmother Fernande to see the property on May 5, 1900, it was a very foggy morning. At first the visibility was poor, but as the fog cleared and the property became visible, she said '*Quel beau lieu,*' which means 'What a beautiful place.' So this is how the property was named Beaulieu Vineyards and Winery."

Over the next few years, Georges planted the BV1 Vineyard, returning to Bordeaux to purchase rootstock.

"They farmed everything biodynamically back then," reported Walter, "and used natural fertilizers in the vineyard and culinary gardens. They raised chickens, ducks, guinea fowl, quail, partridge, woodcocks, turkey, and the like. Though Georges and Fernande passed before I was born, I still remember coming here in the summers and eating fresh trout and crayfish from the local streams. There were livestock, beehives, and fruit trees. All of the vegetables were raised from seedlings in greenhouses and then planted in the garden. The food was amazing, and of course, we had wine with meals. My mother told me that when I was newly born, they placed several drops of wine on my tongue, which is part of the French tradition to dedicate a newborn child to King Henry IV."

Georges and Fernande built a Champagne cellar on the property, and soon their wines started to gain acclaim. By 1908 they had secured a contract to provide sacramental wine to the local Catholic churches, which allowed them to continue to operate during Prohibition from 1919 to 1933. According to Walter, "By 1923 production had increased by four hundred percent, so they purchased the Ewer property across the road." Today this is the site of the BV Winery and visitor's center.

As the years progressed and their original house was lost in a fire, Georges and Fernande moved into another house on the property, which had been built in 1882. They expanded this into the current large mansion seen on the estate today. They also had two children, Hélène and Richard. Hélène, who was Walter's grandmother, married the Marquis de Pins of Gascony and spent half of the year living in France. Hélène had one daughter, Dagmar, who was Walter's mother.

In 1936 Georges approached his son-in-law, the Marquis de Pins, to be his successor at the winery. However, the Marquis declined and suggested to Georges that they travel to France together to look for a new winemaker. While there, they visited the French National Agronomy Institute and were introduced to André Tchelistcheff. André was a Russian aristocrat who had been forced to flee the Kaluga region, south of Moscow, during the Russian Revolution. André studied enology in Czechoslovakia and France and had received job offers from several prestigious wineries but agreed to come to Napa Valley and work at BV in 1938.

When he arrived in Napa, André found that most of the US wineries were in a poor condition with outdated equipment due to Prohibition. He implemented many positive changes including strong sanitary measures in the winery, cold fermentations for whites, and aging in small French oak barrels.

"André also spent much time in the vineyard," said Walter. "Growing up, we called him Uncle André, and he used to send me into the vineyard on horseback to look for vines that had leaf roll. He was very meticulous and wanted to monitor everything in the vineyard to ensure quality."

In fact, it was his excellent palate and focus on quality that allowed André to recognize that the Cabernet Sauvignon from the 1936 vintage of BV1, which he found aging in barrels when he arrived in 1938, was so exceptional that it should be bottled separately. This was the birth of the *Georges de Latour Private Reserve*, a name created by Fernande. To this day, the wine has been made from the grapes of BV1 and is one of the longest continually produced labels in the United States.

With André Tchelistcheff as head winemaker, Beaulieu wines continued to garner much praise, and by the 1940s the wines were served at many White House functions.

"My grandmother Hélène described those days to me," mused Walter.

"Interesting people from around the world came to visit my great-grandparents Georges and Fernande. They loved to entertain. Winston Churchill, Edith Piaf, Rock Hudson, Princess Grace, the Rockefellers, and Élie de Rothschild were among many celebrities who came to Beaulieu where they could enjoy one another's company and converse about world topics."

Fernande de Latour   *André Tchelistcheff*

In 1940 Georges de Latour passed away, but Fernande continued to run the winery with André as her faithful winemaker.

"Growing up," said Walter, "I spent many summers here at Beaulieu. Uncle André visited my parent's residence most Saturday mornings around seven a.m. to have tea with us. He used to tell us the most amazing stories about his life in Russia, Turkey, Bulgaria, and Europe. I remember enjoying the vineyards and gardens with my three sisters Paula, Erica, and Dagmar, named after my mother. They were such happy times."

It was Walter's mother, Dagmar, who planted the row of sycamore trees lining the road to the house with BV1 Vineyards spread out on either side.

"It is a very common custom in France to plant sycamore trees to line roads," reported Walter.

Eventually, in 1969 the family decided to sell the BV Winery operation located on Highway 29 to Heublein, but they retained the vineyards. In 1997 Diageo became the new owner of BV Winery through a series of acquisitions. Today they operate the winery and lease the vineyard from the De Latour offspring.

"We still love coming here on the weekends and during summer," said Walter. "This land is part of our family."

* * *

Tour of BV1

"So you know that BV1 has a total of seventy-eight vineyard acres," said Sam as he drove slowly down the road through the middle of the arching sycamore trees. "But over sixty-one of those acres are planted to Cabernet Sauvignon, while the rest are a mix of Merlot, Petit Verdot, and Malbec. We are going to see the oldest block of Cabernet first."

He parked the truck next to one of the graceful trees with its pale peeling bark, and after clamoring out of the tall vehicle, I followed him into a vineyard block on the left side of the road. It was a tangled glory of tall vines with red-and-yellow leaves that spilled into the middle rows, trailing long strands on the ground. The color of the leaves was so brilliant that I couldn't help but reach out and touch them. The texture felt smooth and cool under my fingers, and I traced the yellow veins that ran through the middle of the brilliantly red leaf. I marveled at how brightly florescent the colors were and thought how much my mother, an artist, would enjoy these glowing hues. However, I knew that regardless of how beautiful they appeared, the red color on the leaves was a sign of virus.

"So you have some leaf roll issues?" I asked, referring to the name of one of the viruses that causes the grape leaves to turn red.

Sam sighed heavily. "Yes, this is an issue in the Napa Valley that we are researching. Many of the older vines have it, and though it doesn't necessarily hurt quality, it can delay ripening and reduce crop load. Despite that, these vines still produced just over three tons of grapes per acre this year."

"They certainly are impressive looking," I said. "How old are they?"

"They were planted in 1989. Obviously the vineyard has been replanted several times since Georges de Latour put in the first vines in the early 1900s."

"So these are no longer on the original rootstock from France?"

"No," responded Sam. "The whole vineyard is now planted on 039-16 rootstock because it is resistant to nematodes, which can be a vector for fanleaf virus. There are several diseases that turn the leaves red and reduce production. One is fanleaf, which we can control with the 039-16 rootstock. Leaf roll and red blotch are two other diseases, but currently there is not a solution to control these."

"So what do you do about it?"

"We monitor the vines as much as possible and may thin the clusters so the vine can ripen the rest of the fruit. We also have a philosophy of replanting every thirty years, with an overall plan to replace five percent per year."

"So these vines will eventually be replaced," I asked, thinking how sad to pull out these huge, red-leafed beauties with their massive, gnarled black trunks and trailing lacy leaves. I know that as a child, I would have wanted to play hide-and-seek among these vines, squatting down in the soil and peering out through the trailing shoots that formed a veil of red-and-yellow Spanish lace.

"Yes," said Sam. "But that is part of the cycle of farming." He turned and gestured across a small dirt road to another block of vines. They were obviously younger due to their smaller trunk size and were void of red leaves. Instead they stood tall and proud in straight rows with a smattering of yellow leaves still clinging to their black shoots.

"Let's check out this next block."

As we walked I was delighted to see a large jackrabbit hopping down a row between the vines. It paused and sat perfectly still for a minute, so that its form was silhouetted against the mountains in the distance. The sun, sliding toward the western sky, highlighted the pink in his long, pointed ears.

"Hey, there's a rabbit," I pointed.

Sam nodded. "Yes, there is a lot of wildlife in the vineyard."

"But not the kangaroos you have jumping around the vines in Australia," I teased, remembering the strange sight of large kangaroos in the vineyards of McClaren Vale and Barossa when I visited there several years ago.

Sam grinned. "No, not here, but as you can see, we have plenty of rabbits and the occasional deer that slips in through the deer fencing."

We continued walking until we reached the vines, and I saw they were planted on VSP trellising with double cordons and spacing of around eight by eight feet.

"These are Petit Verdot," he said. "You'll notice they are on VSP trellis, whereas the older vines we just visited were on Geneva Double Curtain (GDC) trellis. This is because the soil here is more rocky and shallow, which causes the roots to go deeper, so we use VSP. The soil over there is richer because it is closer to the river. GDC trellising is better for richer soils."

Glancing down at the soil, it still seemed to be a rich gray-brown color to me, but Sam pointed at the small rocks and pebbles littering the ground. "This is called gravelly clay. As we move closer to the mountains, we find more gravel and rocks, because it is part of the alluvial fan that flowed down the hillsides into the valley floor here."

He bent down and picked up a rock and handed it to me. It was about two inches long and one inch wide and felt smooth on my palm. I held it up to the sun and was startled to see sparkles of gold and silver flash in the light.

"It sparkles," I said.

"Yes, that is the quartz in the stone," said Sam. "This is classic Rutherford Bench soil."

"Rutherford Dust," I murmured. "Just as rich as gold." I was referring to the local saying about vineyard land in the Rutherford AVA, which was the most expensive land in the Napa Valley and some of the priciest in the world. Trying to buy acreage here was almost impossible, and the price of grapes from this special patch of earth commanded a much higher rate than other areas of the valley. The term "Rutherford Dust" was apparently coined by André Tchelistcheff to describe how special the terroir was in this area.

Old Cabernet Sauvignon Vines in BV1

"You can say that again," said Sam. He turned and ran his hand down the length of a cordon on a Petit Verdot vine. "Even with the pebbly soil, though, we still have to watch the vigor on these vines."

"Why is that?"

"Petit Verdot grows very quickly and puts out many leaves and clusters. Therefore, we have to manage the canopy carefully and thin to one or two clusters per shoot. It is a later-ripening variety, and we want to make sure it is fully ripe when we harvest."

I nodded, thinking of how much I had enjoyed tasting some of the 100 percent Petit Verdot wines from Napa Valley. Though generally used as a blending grape with Cabernet Sauvignon, bottled on its own, Petit Verdot produced an inky purple wine with plush tannins and notes of violet and dark berries.

"Shall we check out a newer block?" asked Sam.

He led the way back across the entry road lined with the majestic sycamore trees and into the grouping of green-leafed vines I had noticed when we entered the property. As we got closer, I could see that a white cardboard protector encircled the base of each vine. This was a common sight in new vineyards.

"So you just planted these vines?" I asked Sam.

"Yes, this is a new block of Cabernet, and we've kept the irrigation on them because they are young. As you know, younger vines need more water."

I nodded, remembering how I had to water the baby vines in my hobby vineyard when I first planted it and also used milk cartons to protect the slender trunks from rabbits or other predators that might be tempted to eat the tender shoots.

"When will you train these to the cordon?" I asked, noting that each vine had around four shoots emerging from the slender trunk.

"We let them grow for two years before staking to the wire."

"But why are there four shoots when you will only have two cordons?"

"Two are for backup, just in case something happens to the others."

Interesting, I thought. "So how often do you irrigate?"

"For these young vines, we irrigate about twice a week during the hot summer months, but once the rains come in the late fall and winter, we will monitor with neutron probes every week to check water levels."

Sam pointed to a series of small sprinklers attached to wires above the vines. "In our more mature blocks, we've actually started experimenting with sprinkling water on the canopy in the summer if the temperature goes over one hundred degrees Fahrenheit. We've discovered that it keeps the grapes at a more constant temperature to promote even ripening."

"Really? That's a fascinating experiment."

"We are doing a lot of experimentation in this vineyard and others as well," said Sam. "Come on, and I'll show you an experimentation block up near the old house."

We climbed back in the truck and drove further into the vineyards toward the De Latour estate, positioned near the foothills of the Mayacamas range. As we approached I could see the stone building of the old Champagne Cellars on the left and Walter's remodeled bunk house and stables on the right. As we drove further into the property, a large barn appeared on the right, and to the left, hidden behind tall oak trees, I could just make out the long peaked roof of the Georges de Latour mansion with the huge expanse of formal gardens in front.

Sam steered the truck past the house and into a small block of vineyards on the other side that bordered the edge of a steep hillside. Parking, we descended and walked to the edge of the block.

"Wow, this is right next to the forest," I said, realizing the last row of vines was only about eight feet from the steeply sloping mountain covered with oak, pine, and madrone trees.

"Don't you have problems with deer eating the grape leaves?"

"Well, we do have deer fencing," said Sam, motioning toward the forest. "It is just hard to see it in the dense undergrowth." Looking closer, I could see a thin stretch of woven wire that composed a deer fence. "However," Sam continued, "we have had to replace a few vines along this stretch near the hillside because the deer still managed to slip through."

I looked at the long row of vines and realized, from the huge thick trunks and wide shaggy cordons, that most were quite old. In a few places, a smaller vine was growing, attesting to a newer planting. The trellis system was also different, looking like a VSP but with the shoots divided into two branching rows.

Experimental Block with New Shoot on Amputated Trunk

"We've created a Modified Lyre trellis system here," said Sam. "These are old vines, and they were planted on a northeast orientation. However, this causes sunburn issues, so to protect the clusters, we modified the trellis system to provide more shade."

"But why were they planted on a northeast orientation?"

Sam shrugged. "Most likely due to practical reasons. As you can see, this is a very narrow strip of land between the buildings and the hillside. If they had oriented it the other way, the rows would be too short, and the tractors would have to turn around multiple times. So instead, they planted it with long rows, and we've modified the trellis system."

"Very clever!"

Sam grinned. "Come on, I'll show you the experimental plot I was telling you about earlier."

We walked across a small dirt road and into a strange-looking group of about fifty vines that appeared to be amputated at the trunk. However, from each sawed-off trunk, there was one slender shoot trained upward on a VSP trellis. They all appeared to have bright green leaves, signaling irrigation at this time of the year.

"What's going on here?" I asked.

Sam bent down to the ground and touched one of the amputated trunks. "As you can see from the width of these trunks, these are old vines, and they had *Eutypa*. *Eutypa* is a fungal disease that may occur if pruning wounds get wet from rain. It causes sections of the vine to die but not the root. Therefore, instead of tearing them out, we decided to experiment and cut them back to start a new shoot."

Admirable, I thought, that BV was so interested in continuous improvement in the vineyard and conducting experiments that could ultimately help the whole industry. Though it was a bit shocking to see the sawed-off trunks of the vines, the fact that they were attempting to let them grow in a new way, rather than just pulling them out as was the more common practice, was heartening.

BV1 Vineyard Specifics

As we continued our tour of BV1, Sam provided more information about the specifics of the vineyard. He explained that the *total vineyard acres* of seventy-eight were comprised of four *varietals*: sixty-one acres of Cabernet Sauvignon, eight acres of Merlot, six acres of Petit Verdot, and three acres of Malbec.

Sam's description of the *soil* matched perfectly with that of the *Rutherford Dust Society,* a group that promotes the special aspects of the AVA:

"The alluvial fan soil is primarily gravelly, sandy and loamy...Deep and well-drained...and dominated by the Franciscan marine sedimentary materials with some volcanic deposits, primarily Bale, Pleasanton, and Yolo Loams (p. 2)."

Elevation for BV1 ranged from 180 feet above sea level closer to the river and extended to almost 300 feet near the foothills. Average summer *temperatures* were 85°F during the day and around 50°F at night. *Rainfall* for the Rutherford AVA averages 38 inches per year.

One interesting aspect of BV1 is that all vines are on 039-16 *rootstock*, which according to Sam, "is resistant to nematodes, which can be a vector for fanleaf virus." He mentioned, however, that an issue with 039-16 *rootstock* is that "it can get overly stressed and can be too vigorous. Therefore, we have to monitor the vineyard carefully."

Clones vary widely by varietal and soil type. Older Cabernet Sauvignon vines are on clones 4, 6, and 7, whereas newer plantings are on 30, 337, 412, and 685. Sam explained that one issue with clone 6 on older vines was shatter, meaning some of the grape berries did not form completely. "Therefore, we are letting one more shoot grow so we can get another cluster. This is part of our experimentation program, and we refer to it as 'rods per vine.'" Clones for other varietals include: Merlot 181, Petit Verdot Sterling 1 and Entav 400, and Malbec 598.

Row orientation for the majority of the vineyard is southwest so the clusters receive enough light and warmth but do not suffer from sunburn. However, small portions have been laid on a northeastern exposure to allow fewer turns for farming equipment. *Spacing* is eight by ten feet for older vines, while newer blocks are planted on eight by eight or four by six spacing. *Trellis systems* are carefully thought-out based on soil type and orientation of the vines. While most new blocks are trained to VSP, older vines in richer soil use Geneva Double Curtain to allow more growth and sprawl, while other portions of the vineyard use a modified Lyre system to prevent sunburn.

Vineyard Specifics for BV1

Total Vineyard Acres	78
Varietals	61 acres Cabernet Sauvignon, 8 Merlot, 6 Petit Verdot, 3 Malbec
Soil	"Rutherford Dust"—primarily composed of gravel, loam, and sand with some gravelly clay closer to the river; well-draining with multiple alluvial fans; some volcanic soil and Franciscan marine deposits"
Elevation	180 to 300 feet above sea level
Average Temperature	Summer temperatures average 85°F for a high and 50°F as a low. Rainfall is usually around 38 inches per year.
Rootstocks	039-16
Clones	Cabernet Sauvignon: 4, 6, 7, 30, 337, 412, and 685; Merlot: 181; Petit Verdot: Sterling 1 and Entav 400; Malbec: 598
Sun Exposure	Primarily southwest, with a few blocks running northeast
Spacing	8 x 10 for older vines; 8 x 8 and 4 x 6 for newer plantings
Trellis Systems	VSP, GDC, and modified Lyre

Farming Practices—Certified Napa Green, Fish Friendly, and Sustainable

"We believe in protecting the environment," said Sam, "and therefore, we have pursued several *certifications*." He paused and ticked them off his fingers. "We are certified Napa Green, Fish Friendly, and

California Sustainable Winegrowing."

"Impressive," I said. "Tell me more about your farming methods."

"For *fertilization* we use a cover crop of mixed beans and grasses, which we sow right after harvest." He pointed at the next row over. "You can just see it coming up over there. When we get more rain, it will be more verdant. We plant cover crops in all rows with new vines, but only every other row with established vines. Later we plow it back into the soil so it can provide nitrogen to sustain the vines. We also add natural compost and will add other organic or nonorganic fertilizers if needed."

"Such as?"

"Well, we do a petiole analysis each year when the vines are at about eighty percent bloom. Based on the results, we may add boron, zinc, or magnesium. This is usually applied as a foliar spray overhead."

For *weed control* the vineyard crew mows beneath the vines and will spray dormant herbicides if necessary. *Canopy management* is done both by hand and mechanically.

"We use machines for hedging and prepruning," said Sam, "but we will hand-prune the rest of the vineyard."

"What about thinning clusters?" I asked.

"We do that by hand. Usually twice before verasion to drop unripe clusters and once after to trim green berries."

"Lots of work."

"Yes, but during harvest we now pick up to sixty percent by machine. We find we are getting almost identical quality, because the machine has an optical sorting mechanism."

"So why not use it on the whole vineyard?"

"Certain trellis systems, such as the GDC, don't allow for mechanical harvesting."

The only *disease* issue is powdery mildew, which is controlled by organic sulfur and DMI fungicides. *Pests* such as birds, gophers, and yellow jackets are not a problem. Sam reported they are using a product called Dust-Off to keep dust mites at bay, and the 039-16 is helping to prevent nematodes. Deer are generally deterred by the deer fencing.

In terms of *irrigation*, all vines have a drip system as well as sprinklers if needed for frost protection, and more recently for experimental cooling in the canopy.

Farming Practices at BV1

Certifications	Napa Green, Fish-Friendly Farming, California Sustainable Winegrowing
Fertilization	Cover crop, natural compost, and other organic and nonorganic fertilizers as needed
Weed Control	Mowing, dormant herbicide sprays
Canopy Management	40 percent harvested by hand, 100 percent pruned by hand; machine for prepruning, hedging, and some harvesting
Disease Control	Organic sulfur and DMI fungicides to prevent powdery mildew
Pest Control	No major issues, using 039-16 rootstocks to control nematodes. Deer fencing for deer.
Irrigation	Drip and sprinklers
Technology	Weather stations, neutron probes, and pressure bombs
Harvest Measurements	Winemaker determines by taste and lab measurements; usually around 26° Brix

"We use neutron probes every week to check water levels," explained Sam, "but this Rutherford soil keeps the rainfall through April, so we usually don't need to start irrigation until summer."

Technology is widely adopted in the vineyard with weather stations, pressure bombs, and neutron probes being used. "We also conduct aerial imagery for vine view in August before harvest," stated Sam. "This helps us gage the condition of the vineyard. In general, we believe in experimentation and the use of new technology for continuous improvement."

Harvest measurements are determined by the winemaking team. Though it depends on the vintage, they generally aim for a Brix of around 26°.

Vineyard Economics of BV1

The *average yield* of the seventy-eight acres in BV1 is 3.2 tons.

"We generally get between two and four tons per acre," said Sam, "depending on the year and the specific block." This calculates to about 250 tons of "Rutherford Dust" grapes harvested in BV1 each year. Interestingly, Sam noted they are starting to switch to the term "pounds per foot" rather than "tons per acre," as it is more accurate and allows for different spacing.

Economic Viability of Vineyard

Average Yield	3.2 tons per acre
Total Average Tons Per Year	250
Costs Per Acre	$6,500
Revenues	Not specified
Economic Health	Quite good

Vineyard work for BV1 is outsourced to Walsh management firm, but Sam oversees more than one hundred full-time workers in the other two thousand acres of vineyards that he manages.

"Octavio is the full-time supervisor here," said Sam, "and he knows the vineyard backward and forward. I stay in very frequent contact with him."

Average *farming costs* are $6,500 per acre, and cost to install a new vineyard ranges from $27,000 to $30,000 per acre. Though a publicly traded company, it is difficult to specify the exact *revenues* of wines produced from BV1. However, as current case production of the Georges de Latour Private Reserve is approximately 8,500 and the retail price per bottle is $120 or $1,440 per case, gross revenues could average around $12 million. According to Sam, the *economic viability* of BV1 is "quite good."

Soul of the Vineyard—"Terroir"

We were driving back toward the gate and Highway 29 when I asked Sam my last few questions.

"If you had to describe this vineyard in one word, what would it be?"

He pulled the truck over to the side of the road and put it in idle. Overhead, the sycamore trees created a beautiful canopy of yellow leaves that shaded us from the sun.

"I know this sounds cheesy," said Sam, turning his head to look at me, "but I would have to use the term 'terroir.'" The sun glinted off his dark glasses, as he turned to gaze out at the vineyard again. "I've worked in vineyards across California, but as André said, 'It takes great dirt to make a great Cabernet,' and Rutherford has that dirt. It is a fantastic site, and there is a strong sense of place here."

"What's the best and worst part of working in this vineyard?" I asked.

"The best part is consistently good quality," Sam said immediately. "The soil is uniform, and we know how it performs. The climate is good, frost is rare, and we generally have a great harvest. In terms of the worst," he paused and turned to look at me again, "well, there really is no bad aspect. It is challenging sometimes to deal with leaf roll and the replanting, but that's part of vineyard work in California."

"What have you learned from the vineyard?"

Party in Gardens of Old De Latour Manor House—1949

He smiled. "We do so much experimentation here, we are always learning. For example, we started pulling leaves around the clusters at prebloom." His voice took on a note of excitement as he continued. "In this way there is more space for the grapes to form because they are not as tightly packed, so we get better flavor development. We are replanting and trying tighter spacing, different trellising, and more mechanization. We are focused on efficiency and quality and trying new ways to be better. It's fun."

As we drove slowly back toward the winery, Sam's words "It's fun" lingered in my mind. It was similar to what Walter had described about the "happy times" he and his sisters had enjoyed playing in the vineyard and working with Uncle André. The property did indeed have a sense of liveliness about it, perhaps not only due to the experimentation taking place, but because of all the interesting people who had visited the property over the years. I could image them eating, drinking, listening to music, dancing, laughing, and arguing late into the night, with the sounds of the revelry wafting out over the vines.

Now, in the late afternoon sun, the vineyard glowed with hues of yellow, red, and orange, seeming to express a hint of lively excitement in a rippling of leaves. It was almost as if the vines waited for the next party, surrounding the historic homestead, protecting it, and providing the economic nourishment of Rutherford Dust.

Signature Wine: Georges de Latour Cabernet Sauvignon

Back at the BV Winery offices, Jeffrey Stambor typed furiously at his computer keyboard.

"Be with you in a minute," he said, glancing over at me quickly as I stood at the doorway of his office. "We're trying to get some wine transferred into barrels. Have a seat."

I walked slowly over to a round table with four chairs situated to the far right of his desk. Two bottles of Georges de Latour 2010 were standing on the table next to two wine glasses. Sitting down, I looked around the office at the many photographs and maps covering the walls. One that caught my attention was an old black-and-white photo of a young André Tchelistcheff standing next to an older woman dressed in a 1940s suit and hat. That must be Fernande, I thought, and wondered what her life was like as a woman managing a winery during that time period.

"So do you want to taste first or see the winery?"

I looked up to see Jeffery rising from his chair. He looked much younger than I had expected, given that he had worked at BV for more than two decades. He wore his black hair cropped short, but with thick eyebrows, a mustache, goatee, and a small ruby earring in one ear, he had a hippy edge about him that was engaging. He was dressed in blue jeans and a long-sleeved red T-shirt, and his movements were quick and energetic.

"The winery, please."

"All right, let's go. Here, you need a safety vest. I see you wore boots. Good."

He tossed me a florescent-green safety vest and started out the door. I followed quickly, having to run a bit to catch up with him.

"So you're a UC Davis grad?" I asked, chasing after him.

"Yes. I studied Plant Science with a specialty in Viticulture." He continued walking quickly down a long hallway and then glanced over at me. "While going to school, I worked part-time at Rutherford Hill Winery and lived in a VW bug in their parking lot for several months."

"A VW bug?"

"Yes." He grinned, and his teeth flashed white under the generous black mustache. "I guess you could say I was a bit of a hippy back then."

"But you started at BV in 1989?"

"Yes," he nodded in agreement, "primarily as the link between the vineyards and the winery. I was fortunate in that André came back as a consultant in 1990, and since I was new and young, they asked me to drive him back and forth to work every day, so I got to know him very well."

"Wow, how fortunate," I said impressed. "What was he like?"

"He was quiet and humble in his older years when I worked with him, but he was never satisfied with the status quo. He was always looking to improve processes in the vineyard and winery." Jeffrey paused outside of a large set of double doors, and looked at me. "I still remember when we brought him to the winery and explained how we were doing things the same as when he left. He was disappointed and said, 'Why? Why not do something different to improve?' He was always striving to do things better, even up to the end when he died in 1994 at the age of ninety-one."

"Ninety-one years old," I said. "What a great legacy he left. And now you are in charge of making the Georges de Latour."

"We call it GDL for short." He flashed a smile and pushed open the double doors. "Here's where we make it. We've created a completely separate production facility to produce the GDL."

Jeffrey Stambor, Winemaker

We entered a large temperature-controlled warehouse filled with conical stainless-steel tanks on legs, six large wooden fermenters, and racks of 225-liter oak barrels stacked high on top of one another. Hoses snaked across the floor, and employees were busy moving them between tanks and barrels.

Jeffrey walked quickly across the room to a computer stand on the far side and consulted a large flat panel that looked like a small television. Glancing at it I could see it was a diagram of all of the tanks, including temperatures, levels, varietals, and other information.

"Very high tech," I commented.

"Yes, it's amazing what we can do in the winery nowadays with

324

technology," said Jeffrey, "but there are still many parts that are completely old school."

"So tell me how you make this famous and historic wine," I asked.

"Well, this past year, harvest started in BV1 on September twelfth and didn't finish until mid-October. Each morning I go out and taste the grapes, and we make decisions on which blocks are ready to pick. The grapes are then transported to this receiving area in the back." He opened a back door and pointed outside to a cement dock behind the winery. "We have tables set up to hand-sort the grapes, which is a lot of hard work."

Jeffrey went on to explain that all blocks are fermented using a combination of natural yeast and five to six commercial yeast strains. Fifty percent of the juice is fermented in stainless steel, 40 percent in new French oak 225-liter barrels, and 10 percent in the large wooden foudres.

"In this way, we can develop a wine with good complexity and different levels of oak integration."

"How are you punching down the caps during fermentation?" I asked, curious about the small barrel fermentation, because it was a relatively new method in Napa Valley and was carried out slightly differently in wineries adopting it.

"The stainless and wooden tanks use traditional pump overs," he said, "and we also perform some punch downs with them." He paused and glanced over at me before flashing a broad smile. "But we have to roll the small barrels."

"Roll the barrels?"

"Yes, come and look at this. Most of these are still undergoing extended maceration, so you can see what it looks like." We walked over to one of the smaller barrels with a plexiglass head on the end of it. Looking inside, I could see it was three-quarters full of small grapes mixed in with macerating wine.

"We roll these barrels several times a day during fermentation to break the caps and integrate the skins and juice," said Jeffrey. "We also have to open the bung to let out some of the CO$_2$ gases that build up. That's referred to as 'burping' the barrel. Now that fermentation is finished, we are letting the wine go through extended maceration."

I looked once more inside the plexiglass window at the tiny purple grapes, so small they reminded me of caviar. They will be pressed and aged next, I thought, and then eventually end up in a fabulous bottle of Georges de Latour. I wondered fleetingly who would be drinking these wine grapes I was staring at now. Would they be consumed with dinner and shared with friends and family? Would these very grapes be part of a festive occasion in the future or be treasured in someone's dark cellar for years?

"So the total maceration usually takes around forty-five to sixty days," Jeffrey was saying.

"That's a long time."

"Yes, and it's different by vintage. We have a house style here we are trying to recreate each year with Georges, but some years Mother Nature gives us riper fruit than others, so total maceration time is always a bit different."

"So when do you decide to press?"

Jeffrey grimaced. "One of the issues I go back and forth over the most is when to press. I am constantly checking the finished ferment daily, both smelling and tasting it." He paused and looked over at me. "In the past this wasn't a problem. We just pressed the wine when it was finished fermenting. Now we like to extend the maceration to improve integration, complexity, and texture, but it takes more attention as we push the limits."

We looked at the large basket press used to gently separate the wine from the must and then walked over to where employees were pumping finished wine into small oak barrels. "We age in one hundred percent-new French oak Medium Toast," said Jeffrey, "for a period of two years and then another year aging in bottle."

"So the wine is really being made with two hundred percent-new oak?" I asked.

"No, it's not two hundred percent because you will remember that fifty percent of it was made in stainless."

"Right, but what is the point of so much oak? That's not the way it was done in the past."

"Yes, but styles evolve, and we don't think the oak dominates the wine. In fact, by fermenting in oak, the flavors are more integrated and the tannins usually a little more approachable early on."

I nodded in agreement, remembering all of the reasons why this method of oak fermentation was being adopted by so many high-end wineries around the world.

Jeffrey went on to explain that malolactic fermentation started naturally during the aging process in barrel and that they were topped monthly for the first two months and then on an eight-week cycle. The wine ages for two years in barrel and is only racked twice before blending, unless during routine tastings they see a need to rack. Blending takes place six months before bottling so the flavors have time to marry.

In terms of fining and filtering, Jeffrey doesn't believe they are always necessary, due to the long aging process GDL undergoes.

"We haven't fined the wine for the past several vintages. I feel strongly about sterile-filtering the wine, as I don't believe that the long-term risk outweighs the reward, especially given that the wine gets a year of bottle age prior to release. My experience has been that after a year, the wines—filtered or not—aren't significantly different."

Back in his office, Jeffrey deftly opened one of the bottles of Georges de Latour Reserve 2010. As I held my glass up to the light, I could see it was opaque with black depths and a glowing ruby rim. On the nose I smelled a touch of coffee, black berry, and spice. On the palate the wine opened up to show some red fruit, with tar, anise, and more complex herbs and spices on the finish. The fine-grained tannins were well integrated (the oak fermentation paid off), and the wine finished with long length and a naturally high acidity due to the cooler vintage of 2010.

"Congratulations," I said. "This is beautiful."

"As you know, 2010 was a cooler vintage, but we managed to get the Brix to around twenty-six degrees before we picked, and I'm really pleased with the natural acidity in this wine."

"So how does the vineyard contribute to this wine?"

Jeffrey's voice was serious as he continued. "BV1 is our original vineyard. It is on the west side and delivers the best fruit. It is the core of GDL, which is a historic wine produced since 1936. It is our heritage and the essence of Rutherford."

"What does BV1 mean to you personally?"

Jeffrey looked thoughtful. "I feel that I grew up in the vineyard professionally. When I first started working here in 1989, I was involved in the replanting of the blocks after *Phylloxera*. I was young and just starting out, and now that I've been working with the vineyard for more than twenty years, I am starting to believe that the more you think you know, the less you know."

"Ah, just like André became at the end," I said, smiling. "Humble."

Jeffrey smiled too, and I thought of John Milton's famous quote from *Paradise Regained*, "The first and wisest of them all professed, to know this only, that he nothing knew." Perhaps that's what vineyards helped to produce after years of toiling in them, the wisdom of humility.

Chapter Fourteen

Lessons Learned from the Vineyards

It was an early evening in July and I was standing in the middle of my small Pinot Noir vineyard in the Petaluma Gap region of the Sonoma Coast AVA. A light breeze had sprung up and the sun was slowing making its way towards the coastal mountain range in the west, where the fog was beginning to pile up like whip cream frosting. The California quail family that lived nearby hopped on the small rock wall that runs around the property, with the father quail leading the way, followed by four small chicks and mother quail bringing up the end.

This was one of my favorite times of the day, when I went to the vineyard to check the progress of the grapes. Verasion was in full swing with half of each cluster filled with glowing red-purple grapes, and the other half still green. This year it was a bountiful crop and I was concerned because my Pinot Noir, which is normally composed of small tight clusters, had decided to grow shoulders. Therefore, with clippers in hand, I was gently cutting off the excess green wings, and pulling some of the leaves which were shading the clusters.

The call of songbirds rang from the large oak tree nearby and the beguiling scent of summer roses, which I planted at the end of each row for both beauty and as an early indicator of powdery mildew, filled the air. During moments like this, the words of Jack London, which I have in a frame on my kitchen wall, filled my mind:

"The air is wine. The grapes, on a score of rolling hills, are red with autumn flame. Across Sonoma Mountain, wisps of sea fog are stealing. The afternoon sun smolders in the drowsy sky. I have everything to make me glad I am alive.... (1913, John Barleycorn)

Soon I was enveloped in the deep sense of meditative peace that often comes over me when working amidst the vines. Time seemed to disappear and my only focus was the leaves and grapes. In fact, one evening I was so engrossed in leaf pulling that I didn't pay attention to the hose I thought I was standing on. Though the vineyard has an irrigation system, my husband had also dragged a hose down several years ago and left it in one of the rows. Rocking back and forth on the hose with my flip flops, I was feeling a small sense of irritation that he had left it there.

Then I glanced down and screamed. Instead of a hose, I was standing on a three foot black and white striped king snake. Leaping about two feet in the air, I jumped off the snake and it immediately slithered down a nearby gopher hole. After taking several deep breaths, I began to laugh as I realized the snake was probably more scared than I was. Besides I was quite happy to have a king snake in my vineyard, because they eat gophers, of which there were plenty.

There had been other unexpected incidents in the vineyard over the past decade. For example, one year I encountered a yellow jacket nest and was stung repeatedly. There have been several times when I've had to rescue and release small birds that became tangled in the bird netting we put up each year to protect the grapes. Another year I returned from an overseas trip to discover a family of raccoons had eaten the whole crop. They were clever enough to slip under the bird netting.

There were also several challenging years in the beginning when powdery mildew ravaged huge portions of the crop. The third year it happened, I remember sitting down on the ground in the vineyard and crying. I felt like giving up, but eventually I found someone who told me to spray the vines with stylet oil in the spring before bud break and that usually prevents or at least cutback on outbreaks.

In the beginning I tried to farm using organic practices but gave up when pulling and hoeing weeds resulted in painful backaches. The native California grasses would grow so tall and dense around the vines that it would take hours to remove them. My husband was so worried about me, he took matters into his own hands and sprayed the vineyard with Round-Up to kill the weeds. So now we follow sustainable farming practices (which allow Round-up) and try to limit the use of non-organic products as much as possible.

One of the interesting situations of tending a vineyard is the myriad of advice received from others. Over the years as I've encountered various problems, I've hired different consultants, read a lot of books, talked with other vineyards owners, and received many suggestions that were often contrary to one another. Some of the advice worked, and some did not.

Therefore it is partially for this reason, that I was interested in exploring the ten famous vineyards described in this book. I was curious to see what practices they were using to achieve such great success, and what others, including me, could learn from them.

Harvesting Lessons Learned

Trying to whittle down the hundreds of insights, vineyard tips, and nuggets of wisdom provided by the vineyard managers, winemakers, and others who so kindly agreed to grace these pages is humbling. However by using a process of qualitative research called thematic coding, it is possible to identify major concepts and ideas that came up multiple times, and therefore can be deemed to be important lessons learned. They also highlight some of the concepts that are significant at this time in Napa and Sonoma viticulture, and illustrate how California is both similar and different to other famous winegrowing regions of the world.

Therefore, after using this methodology, I was able to identify eight practices and concepts that were evident in the majority of the vineyards. Though there are probably other lessons that can also be gleaned from this process, these seemed to be the most significant.

Lesson 1: The Importance of Matching Climate and Soil to Cultivar

Countless viticulture books describe how important it is to consider and match climate and soil to cultivar (grape varietal, clone and rootstock). This is a concept that has been implemented very well in each of the ten vineyards, though perhaps not from the beginning. At Hirsch, David planted both Riesling and Pinot Noir to start, but later grafted the Riesling to Pinot Noir when he recognized how well matched it was to the climate and site. He also changed the rootstock over the years as he grew to recognize the diverse soil in different parts of the vineyard, which required different rootstock. Likewise Jan at Stagecoach had a similar experience. He discovered over the years that moving varieties such as Malbec, Sauvignon Blanc and Cabernet Franc to slopes that received less direct sun produced better results.

In other vineyards the match of cultivar to climate and soil was part of the original vision. For example, at Diamond Creek, Al set out to find a place in California to grow mountain Cabernet Sauvignon and happened to find the perfect climate but with three unique soil structures each producing a distinctive style of wine. Likewise Ambassador James at Hanzell dreamed of planting a Burgundian vineyard in the foothills above Sonoma and succeeded in finding an ideal climate and soil for Chardonnay and Pinot Noir. At Stag's Leap, Warren spent several years seeking the perfect spot to grown Merlot and Cabernet Sauvignon until the fateful day he tasted Nathan Fay's wine and knew he needed to buy the land next door.

Many of the historic vineyards such as Monte Rosso, To Kalon, Seghesio, BV, and Bacigalupi planted what were deemed to be the most appropriate varieties for their day, and then modified if needed. This concept of matching climate, soil, and cultivar is incredibly important in achieving high quality grapes, and is one that California has had less time to master than the great vineyards of Europe, many of which were established by the Romans. Yet California has still been very successful, given the fact that the Spaniards didn't plant the first vineyard until the 1760's when they established the San Diego mission, and the first premium winery, Buena Vista, was not established until 1857 in Sonoma.

Lesson 2: Freedom to Experiment

Another clear theme that ran through the interviews was experimentation. Many of the vineyard managers described projects they were working on in terms of new types of pruning, rootstock, trellising, cover crops, irrigation methods, canopy management, and grafting over to new varietals. An excellent example is at BV where Sam has established a special experimental block to grow new shoots on old vines that have suffered by *Eutypa*.

This pursuit of continuous improvement and innovation in the vineyard is inspiring because it doesn't happen in every wine growing region of the world. In fact, in many European vineyards strict regulations don't allow for some of these types of changes, especially experimentation with different varietals, trellising and irrigation. This freedom to experiment and innovate is a hallmark of vineyards in Napa and Sonoma, and reflects the spirit of California viticulture in general.

Lesson 3: Sustainable Farming Practices and Respect for Wildlife

Another lesson to be gleaned is that all of the vineyards are using sustainable farming methods, or biodynamic approaches as in the case of Hirsch Vineyards. Their philosophy for doing so is a desire to use natural and organic products as much as possible, but to have the option of employing non-organic substances, such as Round-up, if needed. This illustrates a strong respect for the natural processes of the land, coupled with a practical need to achieve economic success in the vineyard. Indeed these are the basic tenants of the California Sustainable Winegrowing Alliance (CSWA), to be environmental friendly, use equitable social practices towards workers and community, and to be economically viable.

The number of different environmental certifications is also commendable. In addition to the CSWA certification, some of the vineyards are also certified in Fish Friendly Farming, Napa Green and ISO1401. Though not all of the vineyards are certified in sustainable winegrowing, they have adopted the practices because they believe they are useful in both protecting the environment for future generations and in producing higher quality grapes.

One of the interesting results of their efforts is the wide range of wildlife witnessed in these California vineyards. For example, in my visits I saw quail, hawks, turkeys, rabbits, and a coyote, and heard stories about snakes, eagles, owls, bobcats, and mountain lions. Probably some of the most unusual reports were wild boar at Seghesio's Home Ranch Vineyard and bear sightings at Stagecoach Vineyards.

Many of the vineyards had also set up owl boxes, raptor perches, and planted flowering shrubs and grasses to attract beneficial insects. They are taking active steps to encourage wildlife in the vineyard. These efforts help to re-establish the natural patterns of predators hunting prey, which in turn supports the health of the vineyard.

Other sustainable efforts include a focus on water conservation and the use of cover crop to prevent erosion. Both actions are good for the environment and for encouraging grape quality, because less water can reduce vigor in vines, and cover crop can be disked back into the soil to provide natural fertilizer.

Lesson 4: Balance between Technology and Respect for Mother Nature

The theme of Mother Nature verses technology came up in every vineyard, but in the end it seemed as if there was a balance and respect for both. Most of the vineyard managers described issues with frost, rain, heat, or other natural occurrence over which they had little control. Some accepted it with grace, whereas others fought back with technology. Some did both. Perhaps the most clear example of this was at Seghesio where the father, Jim, described that after years of working in the vineyard he accepted the whims of Mother Nature, whereas son, Ned, admitted to trying to "best her." Though they are utilizing technology with weather stations and digital imagery, Jim discovered that some traditional methods work just as well. For example he found that climbing to the top of the hill to look at the vineyard leave color from a distance worked as well as NDVI mapping with airplanes, and using willow switches to trim vines worked better than mechanical hedgers.

The issue of smaller family vineyard verses larger corporate vineyard also seemed to be at play here. For example, family owned vineyards such as Bacigalupi and Hirsch, use very little technology. Instead they relied on the traditional methods of observation. On the other hand, the larger vineyards owned by publicly traded corporations, such as To Kalon, Stag's Leap and BV, utilized most of the new vineyard technology. This could be due to the fact that they have more financial resources to do so, and/or stockholders expect positive returns so every effort is made to control Mother Nature with technology if possible.

Some experts predict that in the future it will be possible to monitor every vine with technology and respond quickly to any Mother Nature issues, such as lack of moisture, frost, too much heat, rain, disease and pests. Though this may happen in large scale agriculture projects, it was apparent that all of the vineyard managers tending these famous vineyards recognize that some aspects of nature can't be controlled. A few also acknowledge that in some cases, nature serves as a natural quality control mechanism by reducing crop load. For example, virus was described as a problem in some vineyards, whereas others admitted that it resulted in fewer clusters, but with grapes with higher natural acidity and good phenolics.

Lesson 5: Admiration for Work Ethic and Wisdom of Mexican Workforce

A clear message that came through in the majority of the chapters was a strong respect, admiration, and dependence on the Mexican workforce in California vineyards. Three of the vineyards, Hirsch, Hanzell and Stagecoach, employed a full-time Mexican-American vineyard manager, whereas the others hired Mexican-American men and women as vineyard workers and foreman, and brought in immigrant workers from Mexico during harvest. The respect and high regard for the skill set of these workers was obvious.

The special wisdom many of these workers have regarding the land and growing vines is regarded by many experts as a competitive advantage for the California wine industry. Though the early history of Mexican immigrant workers in California was not that positive due to often poor and unsafe working conditions, today that has changed. Because of the efforts of leaders such as Cesar Chavez, who fought and won progressive agriculture worker concessions, the industry as a whole is better.

The new question is whether or not these talented employees want to continue to work in California vineyards. Many are already seeking jobs in other industries where they do not have to spend long hours outdoors. Because of this, some vineyards have converted to mechanical vineyard operations, both to prepare for that possibility and to reduce costs in some cases. At the same time, many high-end wineries prefer to have their grapes hand-harvested. The question of how long this can continue is unanswered.

On the brighter side, some of the second and third generation descendants of Mexican-American vineyard workers now own their own vineyards and operate successful wineries. With their wisdom and strong work ethic, some will most likely create the future great vineyards of California.

Lesson 6: Education and Mentoring

Another interesting learning is that the majority of the vineyard managers and winemakers in these ten famous vineyards have a college degree. Though not all degrees are in agriculture and vineyard management, they have obtained additional training in viticulture and winemaking through online courses, workshops and self-study.

The gift of a mentor is also mentioned several times as an important aspect of education. For example, Everardo at Hirsch Vineyards described how his father and uncles taught and mentored him. Likewise Kirk at Stag's Leap was mentored by Warren Winiarski and Jeffrey at BV was fortunate enough to work several years with the great Andre Tchelistcheff before he passed. This theme of education and mentorship for vineyard management and winemaking appears as an important support mechanism for a successful operation.

Lesson 7: Reverence for History and Old Vines

Many of the vineyards described in this book have historical roots, which contribute to their reputation as a great vineyard. For example, To Kalon (1868), Monte Rosso (1880), and Seghesio (1895) were established before the turn of the century, with Monte Rosso still producing fruit from some its ancient zinfandel vines. Stag's Leap (1970) and Bacigalupi (1964) Vineyards both have historical relevance because they contributed grapes that went into wines that won at the Judgment of Paris tasting.

The other vineyards established in the 1900's, also have a strong connection to history. BV (1900) is credited with helping to resurrect the California wine industry after Prohibition with the vision of the great Andre Tchelistcheff. Hanzell Vineyard (1953) is home to the oldest Pinot Noir and continuously producing Chardonnay vines in North America. Hirsch (1980) is the oldest premium Pinot Noir vineyard on the Sonoma Coast, and Diamond Creek (1968) is the first exclusively Cabernet Sauvignon estate vineyard and pioneered the concept of micro-climates on Diamond Mountain. Even Stagecoach (1996), the relative newcomer, has researched its historical roots to find Black Bart and the old stagecoach route, which they have embraced as part of their clever branding strategy.

Another reference to history is the mention of Native American artifacts. Both Boots at Diamond Creek and Jasmine at Hirsch described finding arrowheads on the property, and expressed the awe and reverence they felt for the original populations who lived on the land.

The issue of vine age was raised repeatedly throughout the chapters, with some vineyards attempting to take whatever actions were needed to preserve their heritage vines. Though there are many opinions regarding what vine age is optimal for producing great wines, most of the vineyard managers who were responsible for tending the ancient beauties recognized the need to preserve them as part of California's history.

Lesson 8: Passion, Vision & Humility

Finally, probably one of the most important aspect of success for these vineyards is the clear passion, vision, and humility of the people who work in them. Every person interviewed expressed a strong and abiding passion for their work. Though it is frequently said that the wine industry has passionate people working in it, this was clearly evident in these ten vineyards. The vineyard managers enjoyed working outdoors, relished the art of tending and sculpting the vine, and appreciated the intellectual challenge of working with Mother Nature. The winemakers crafting wines from the grapes of these vineyards all expressed a clear passion and knowledge for the vine, and a strong recognition that a great wine starts in the vineyard.

Vision was another theme that was apparent as a key for success. Many of the vineyard managers and owners had to struggle to overcome hardships in order for the vineyard to succeed. The older vineyards suffered through Prohibition years, whereas other vineyards managed to stay viable during economic downturns, changes in ownership, and different farming philosophies. Jan at Stagecoach had to cling to his vision of a great vineyard even when he thought there was no water and vandals torched his equipment. In all these cases a strong and abiding vision of producing high quality grapes was needed.

Finally, an element of humility crept into these pages. This surprised me because I wasn't expecting it, but many of the vineyard managers and winemakers – especially those with many years of experience – admitted that working in the vineyard made them humble. They recognize that even with all of the science and technology to focus on quality results, that there are some things they don't understand. This humility however, doesn't impact their passion and vision, but instead compliments it with a special grace that allows them to accept what they do not know and to recognize there are other powers and factors at work. It gives them a calming wisdom to continue working, and to become great mentors for the next generation who answer the call of the vine.

Epilogue - Full Circle

As fate would have it, several weeks after I finished visiting the last of the ten famous vineyards of Napa and Sonoma, I found myself back in Burgundy standing in front of the world-renowned Romanée-Conti Vineyard. This time I was with a group of twenty of my California wine students, escorting them on a class tour of French wine regions. It was a gray blustery day in late May with a threat of rain in the forecast. Many of the students were snapping pictures with cell phones and cameras, and standing next to the famous stone cross.

"Why don't we have vineyards like this in California with rock walls and statues?" asked one student.

There it was. The same question I had asked myself several years ago when I last visited this spot. I was just getting ready to answer that we do have many famous vineyards in California when I was interrupted by a loud shout.

"Group photo! Everyone gather round the cross for a group photo."

Someone grabbed my arm and I was dragged into the photo while our French tour guide was begged to take several class shots. We posed and primped and took multiple goofy shots before the wind rose up bringing with it a smattering of cold rain drops. The few who were wise enough to bring umbrellas hoisted them overhead and we all raced back down the road to the square in the tiny village of Vosne-Romanée where the bus was parked.

As we drove slowly away, several hands shot up from the back seats. "Why do they carve the names of the vineyard in stone and put up crosses?" someone yelled, and then "Why don't we have vineyards like that in California?"

Reaching for the bus microphone, I stood up and said, "But we do have some very famous vineyards in California, and a few of them now have signs with the name of the vineyard. Since we have a little time before we get back to our hotel, let me tell you some stories about famous California vineyards ..."

Appendix

Other Famous Vineyards in California

Following is a list of nominations of other famous vineyards in California. These are from regional winery and vineyard associations I contacted requesting they identify famous vineyards in their region. The same criteria were given: please list three to five of the most historic, unique, and/or high quality vineyards (producing large number of award winning wines) in your region.

Regional Association	Vineyard Nominations (67)
Lake County Winegrape Commission	*Historic vineyards:* Langtry, Kendall Jackson, Holdenried Vineyards, Dorn Vineyards, and Luchsinger Vineyards *Very Unique:* Shannon Ridge Vineyards (30% slope) uses sheep to leaf pull (not too many a re doing that?) Dorn Vineyard (they harvest the old Chilean way with buckets) Obsidian Ridge- 2680 feet - 1/2 mile up. (Actually has a wine called 1/2 mile) *Vineyard that consistently produces a high number of award winning wines.* Andy Beckstoffer's Red Hills Vineyards, Shannon Ridge Vineyards', Langtry, and Jed Steele's vineyards

Lodi Winegrape Commission	Jessie's Grove Royal-Tee Vineyard (over 120 years old). Marian's Vineyard (planted 1901) Dogtown Vineyard Noma Ranch Vineyard Kevin Soucie Vineyard (multiple award winner) Silvaspoons Vineyards (unique in Portuguese varieties, all winning a string of awards) Bokisch Ranches (pioneered Spanish varieties in Lodi)
Mendocino Winegrape & Wine Commission	Navarro Vineyards (Been in businesses since 70s with great reputation for quality) Roederer Vineyards (famous for sparkling wines) Bonterra Vineyard (very successful organically grown brand with consistent quality) Parducci Vineyards (oldest winery in Mendocino). Dark Horse Vineyard (unique as leader in biodynamics; high quality)
Monterey County Vintners & Growers Association	Chalone (historic and famous for Pinot Noir) Wente (historic, famous clone) Talbott Morgan J Lohr
Paso Robles Wine Country Alliance	Steinbeck Vineyards & Winery Dusi Vineyard Pesenti Adelaida's HMR Vineyard James Berry Vineyard Denner Vineyards J. Lohr's Hilltop Vineyard

344

San Diego County Vintners Association	Orfila Vineyards (Located in San Pasqual Valley, started by former Ambassador Alejandro Orfila)
	Shadow Mountain Vineyards (Located in the highest elevations of the County, producing excellent wines)
	Bernardo Vineyards (one of the oldest in San Diego)
San Luis Obispo Vintners & Growers Association	Edna Valley Vineyard
	Larner Vineyard
	Tolosa Vineyards
	Talley Vineyard
Santa Barbara County Vintners Association	Bien Nacido Vineyard – Santa Maria Valley AVA
	Byron's Vineyard – Santa Maria Valley AVA
	Zaca Mesa "Black Bear Block" – Santa Ynez Valley AVA (First Syrah planted in Santa Barbara County – 79)
	Sanford & Benedict Vineyard – Sta. Rita Hills AVA (planted in '71. Known for both Pinot Noir and Bordeaux varietals)
	Firestone Vineyard – Santa Ynez AVA (First estate vineyard/winery operation planted '72 first crush '75)
	Beckmen's "Purisima Mountain Vineyard" in Ballard Canyon (Organic/biodynamic; high praise for its Rhône-based varieties)
	Gypsy Canyon Vineyard – Sta. Rita Hills AVA (Historic and Unique with *Mission* grapes growing, and direct lineage to *Mission La Concepcion Purisima*, and its vitis vinifera fruit brought to California by the Spanish padres, truly California's heritage grape)

Santa Cruz Mountains Winegrowers Association	Ridge Mount Eden David Bruce Woodside Hallcrest Vine Hill Paul Mason/Martin Ray Burrell School
Sierra Foothills (including Sierra Foothills AVA, Amador & El Dorado counties)	Deaver Vineyards (famous old vine zinfandels) Cooper Vineyards (many awards over the years, best barbera in CA) Eshen Vineyard Grandpère Vineyard Fosatti Vineyard
Temecula Valley Winegrowers Association	*Historic* Hart Vineyards (Syrah vineyard dating back to 1974) Brookside Winery Vineyards (planted back in the mid 1960's) *Very Unique* Baily Vineyard (grows only Bordeaux varietals) Cougar Vineyard (grows only Italian varietals) *Top Quality Award Winning* Leoness Vineyards South Coast Winery & Vineyards

References & Photo Credits by Chapter

Chapter One: The Soul of the Vineyard
NVV. (2014) Napa Valley Appellations. Available at: http://napavintners.com/
SCV. (2014) Sonoma County Appellations. Available at: http://www.sonomawine.com/
Photo/Map Credits: AVA and Vineyard Maps designed by Michelle Drewien of Zango
    Creative, Seattle, WA.

Chapter Two: Viticulture 101
Adler, J. & Weingarten, T. (2005). The Taste of the Earth. *Newsweek*, Vol. 145, Issue, 9,
    p. 54.
Carey, V. A., Archer, E. & Saayman, D. (2000). Natural terroir units: What are they?
    How can they help the wine farmer? *Wynboer*. Available at:
    http://www.wynboer.co.za/recentarticles/0202terroir.php3.
Dolan, P. (2011). Personal Interview. Dark Horse Ranch, Ukiah, CA
Field, S. (2004). Minervois La Liviniere: A blueprint for the future of the Languedoc?
    Journal of Wine Research, Vol. 15, Issue 3, p. 227.
Gade, D.W. (2004). Tradition, Territory, and Terroir in French Viniculture: Cassis,
    France, and Appellation Controlee. *Annals of the Association of American
    Geographers*, Vol. 94, Issue 4, P. 848.
Gladtones, J. (1992). Viticulture and Environment. Adelaide, Australia: Winetitles.
Goode, J. (2003). Terroir: muddy thinking about the soil. *The Wine Anorak*. Available
    at: http://www.wineanorak.com/index.htm
Johnson, H. (1994*). The World Atlas of Wine*. London: Simon & Schuster; 4th edition.
Robinson, J. (1999). *The Oxford Companion to Wine*. Oxford: Oxford University Press,
    2nd edition.
Seguin, G. (1986). Terroirs and pedology of wine growing. *Experientia*, Vol. 42, P. 861.
Smart, R. (1991). *Sunlight Into Wine; A Handbook for Wine Grape Canopy
    Arrangement*. Adelaide, Australia: Winetitles.
Terroir-France (2013). Guide to French Wine Appellations. Available at:
    http://www.terroir-france.com/wine/
Thach, L. & D'emilio (2008). *How to Launch Your Wine Career*. San Francisco: Wine
    Appreciation Guild.
UC Davis. (2012). Viticulture Information: Powdery Mildew. *UC Davis Integrated
    Viticulture*. Available at:
    http://iv.ucdavis.edu/Viticultural_Information/?uid=66&ds=351.
US Legal. (2012). American Viticulture Areas Law and Legal Definition. Available at:
    http://definitions.uslegal.com/a/american-viticultural-areas/
WSU (2009). Grape Diseases. *WSU Viticulture & Enology Research and Extension*.
    Available at: http://wine.wsu.edu/research-extension/plant-health/grape-
    diseases/
WSU (2013). Red blotch verses leafroll. A New Virus Disease Poses Additional Threat
    to Viticulture in Washington. *WSU Viticulture & Enology Research and
    Extension*. Available at:
    http://wawgg.org/files/documents/2013_June_WSU_and_Red_Botch_Fact_She
    et.pdf?PHPSESSID=505b33bd008d6d35dd8ab91416967f2f
Photo Credits: Robert Mondavi photographic archives, Jim's Supply.com, Santa Rosa JC,
    Liz Thach

Chapter Three: Monte Rosso Vineyard

Heald, E & R, (2007). Sonoma Valley's Legendary Monte Rosso Vineyard: Acclaimed Winemaker Ed Sbragia Explains Its Allure. *Appellation America.com*. Available at: http://wine.appellationamerica.com/wine-review/413/Monte-Rosso-Vineyard.html.

Historic Vineyard Society, 2011. Monte Rosso Vineyard. Available at: http://www.historicvineyardsociety.org/registry/.

Kleinschmidt, J. (2009). Prime Property: Winemakers line up for Monte Rosso grapes. *Palm Springs Life*. Oct. 2009. Available at: http://www.palmspringslife.com/Palm-Springs-Life/October-2009/Prime-Property/.

Martini, L. M. (2012). "Our Vineyards – Louis M. Martini." Available at: www.louismartini.com/about-us/our-vineyards.html.

Martini, L.M. (2012). "Monte Rosso Vineyard Fact Sheet."

Mazzetto, F., Calcante, A., Mena, A., and Sacco, P. (2011). "Test of Ground-Sensing Devices for Monitoring Canopy Vigour and Downy Mildew Presence in vineyards: First Qualitative Results." *Journal of Agriculture Engineering*, (2011), **2**, 1-9. Available at: http://www.google.com/search?client=safari&rls=en&q=viticulture,+ndvi+maps&ie=UTF-8&oe=UTF-8.

NASS. (2014). *Final Grape Crush Report – 2013 Crop*. USDA Publications. Available at: http://www.nass.usda.gov/Statistics_by_State/California/Publications/Grape_Crush/.

PRWeb, (2012). "Monte Rosso Vineyard of Napa Valley's Louis Martini Winery Wins 2012 Vineyard of the Year Award." *PR Web*, August 2012. Available at: http://www.prweb.com/releases/best-wineries/napa-valley/prweb9783385.htm.

Wine Business Monthly, (2013). "How the UC Davis Gubler-Thomas Powdery Mildew Model and McCrometer CONNECT Help Wine Growers Combat Grapevine Powdery Mildew." *Wine Business Monthly*, January 2013. Available at: http://www.winebusiness.com/suppliernews/?go=getSupplierNewsArticle&dataId=110222.

Photo Credits: Courtesy of E&J Gallo's photographic archives or by Liz Thach

Chapter Four: Tokalon

Constellation Academy of Wine. (2007). "H.W.Crabb: The Man of To Kalon." *Blog of Constellation Academy of Wine Blog*. Available at: http://www.academyofwine.com/our-blog/2007/8/28/h-w-crabb-the-man-of-to-kalon.html

Constellation Brands. (2012). *Annual Report*. Available at: http://www.cbrands.com/home.

Lukacs, P. (2000). *American Vintage: The Rise and Fall of American Wine*. NY: Houghton-Mifflin.

Mondavi, R. (1998). *Harvests of Joy: How the Good Life Became Great Business*. Fl: Harcourt Brace.

Robert Mondavi Winery. (2013). Website of Robert Mondavi Winery. Available at: http://www.robertmondavi.com/.

Photo Credits: Courtesy of Robert Mondavi Winery photographic archives or by Liz Thach

Chapter Five: Seghesio

Boitano, J.J. *Alla Corrente - The Seghesio Story*. Available at: http://www.ilcenacolosf.org/images/Seghesio_Story.pdf

Hitch, D. (2009). Italian Swiss Colony is a brand which rings a resounding bell. *Mercer Island Reporter*, July 7, 2009. Available at: http://www.mi-reporter.com/lifestyle/50146512.html.

Pinney. T. (1989). *History of Wine in America: From the Beginnings to Prohibition.* Berkeley: UC-Press.

Seghesio (2014). *Seghesio Story.* Available at: http://www.seghesio.com.

Photo Credits: Courtesy of Seghesio Winery photographic archives or by Liz Thach

Chapter Six: Stag's Leap

Altria, Inc. (2014). Altria 2013 Annual Report. Available at: https://materials.proxyvote.com/Approved/02209S/20130325/AR_161589.PDF

Asimov, E. (2007). "Stag's Leap Sold for $185 Million." The New York Times, July 31, 2007.

Department of Fish & Wildlife. (2013). List of California Endangered Species - 2013. Available at: http://www.dfg.ca.gov/biogeodata/cnddb/pdfs/TEAnimals.pdf

Lukacs, P. (2005). The Great Wines of America – The Top Forty Vintners, Vineyards, and Vintages. NY: Norton.

Swinchatt, J. & Howell, D.G. (2004). The Winemaker's Dance: Exploring Terroir in the Napa Valley. Berkeley: University of California Press.

Taber, G. M. (2005). Judgment of Paris – California Vs. France and The Historic 1976 Paris Tasting That Revolutionized Wine. NY: Simon & Schuster Inc.

Photo Credits: Courtesy of Stag's Leap Wine Cellars photographic archives or by Liz Thach

Chapter Seven: Hirsch

Gruver, J., Weil, R.R., White, C., & Lawley, Y. (2014). "Radishes – A new cover crop for Organic Farming Systems." *Extension.* Available at: http://www.extension.org/pages/64400/radishes-a-new-cover-crop-for-organic-farming-systems#.U7bTl1x6elI

Hansen, H. & Miller, J. (1962). *Wild Oats in Eden: Sonoma County in the 19th Century.* Santa Rosa, California; Unknown Publisher.

Heimoff, S. (2012). "New California AVA Approved – Fort Ross-Seaview gets the Federal green light." *Wine Enthusiast.* Available at: http://www.winemag.com/Web-2011/New-California-AVA-Approved/

Hirsch Vineyards (2013). "The Site." *Hirsch Vineyard Website.* Available at: http://www.hirschvineyards.com/The-Site

Molesworth, J. (2013). "A Stop at Hirsch Vineyards." *Wine Spectator.* Available at: http://www.winespectator.com/blogs/show/id/48507

Moore, H. (1997). "Rudolf Steiner: A Biographical Introduction for Farmers." *Biodynamic Association.* Available at: https://www.biodynamics.com/rudolf-steiner-biographical-introduction

Moyer, M. (2012). "Grapevine Powdery Mildew." *Washington State University Extension.* Available at: http://wine.wsu.edu/research-extension/files/2011/02/2012-EastPMWhitePaper.pdf

Reeve, Jennifer R.; Lynne Carpenter-Boggs, John P. Reganold, Alan L. York, Glenn McGourty and Leo P. McCloskey (December 1, 2005)."Soil and Winegrape Quality in Biodynamically and Organically Managed Vineyards". *American Journal of Enology and Viticulture* 56 (4): 367–376.

UC Davis (2014). "Eutypa Dieback." *University of California Viticulture Information.* Available at: http://iv.ucdavis.edu/Viticultural_Information/?uid=64&ds=351.

YT&T. (2014). Scientists say ancient dumps helped bay. *History Matters – Yesterday, Today & Tomorrow.* Available at: http://www.yttwebzine.com/87267

Photo Credits: Courtesy of Hirsch Vineyard photo archives and L. Thach

Chapter Eight: Stagecoach

Baumgartner, K. & Rizzo, D.M. (2000). Oak Root Fungus on Grapevines. *Practical Winery & Vineyard Journal,* April/May 2000. Available at http://www.practicalwinery.com/marapr00/oakrootfungus.htm

Blackbart.com (2013). Black Bart, California's Infamous Stage Robber. *BlackBart.com.* Available at: http://www.blackbart.com.

Coodley, L. & Mathews, C. (2013). The Haunting of Soda Springs Available at: http://laurencoodley.com/sodasprings.html

Golino, D. (2008). Sources of Cabernet Sauvignon Clones. *UC Davis Integrated Viticulture Online.* Available at: http://stream.ucanr.org/cabernet/Golino/index.htm

Jensen, P. (2012). Death of Monticello was a heartbreaker. *Napa Valley Register*, August 2012. Available at: http://napavalleyregister.com/news/local/death-of-monticello-was-a-heartbreaker/article_5a2687d8-ea83-11e1-a694-0019bb2963f4.html

Johnson, L. F., et al. (2003). Mapping vineyard leaf area with multispectral satellite imagery. *Computers and electronics in agriculture* 38.1 (2003): 33-44.

Lindblom, J. (2012). The rich history of hotels in the Napa Valley. *The Weekly Calistogan, Napa Valley Register,* March 2012. Available at: http://napavalleyregister.com/calistogan/lifestyles/the-rich-history-of-hotels-in-the-napa-valley/article_09fc0366-6336-11e1-96c4-001871e3ce6c.html

Paul Hobbs (2013). Technical winemaking notes for Paul Hobbs Cabernet Sauvignon Stagecoach Vineyards 2007. Available at: http://www.paulhobbswinery.com/

Stagecoachvineyards.com. (2014). Vineyards and People of Stagecoach. Available at: http://www.stagecoachvineyard.com

Swinchatt, J. & Howell, D.G. (2004). *The Winemaker's Dance: Exploring Terroir in the Napa Valley.* Berkeley: University of California Press.

Photo Credits: Courtesy of Stagecoach Vineyards Photographic Archives and Liz Thach

Chapter Nine: Bacigalupi

John Tyler Wines Website. Available at: http://johntylerwines.com/.

NASS. (2014). *Final Grape Crush Report – 2013 Crop*. USDA Publications. Available at: http://www.nass.usda.gov/Statistics_by_State/California/Publications/Grape_Crush/.

Taber, G.M. (2005). Judgment of Paris: California Vs. France and the Historic 1976 Paris Tasting That Revolutionized Wine. NY: Scribner.

Thach, L. (2013) "A Legendary Vineyard: Source of Chardonnay That Helped Win Judgment of Paris is Still Thriving." *WineBusiness.com*, June 4, 2013. Available at: http://www.winebusiness.com/news/?go=getArticle&dataid=117399.

Wente, K. (2010). Wente Vineyards, the birthplace of Chardonnay. Karl Wente Blog. Available at: http://www.wentevineyards.com/blog/wente-the-birthplace-of-Chardonnay/

Wine Library of Sonoma County, (2010). "Oral History of Charles & Helen Bacigalupi." Tape-recorded and prepared by Vivienne Sosnowski, Nov. 18, 2010.

Photo Credits: Courtesy of Bacigalupi Vineyard photographic archives or by Liz Thach

Chapter Ten: Hanzell

Bogart, K. (2005). "Measuring winegrape water status using a pressure chamber." Available at: http://pmsinstrument.com/Kay%20Bogart%20Article%2001132005.pdf

Hanzell Vineyards Website. (2012). "Vision, Place, and Winemaking of Hanzell Vineyards." Available at: http://hanzell.com/

USDA. (1972). Soil Survey of Sonoma County. University of Michigan Press: U.S. Dept. of Agriculture

Photo Credits: Courtesy of Hanzell Vineyard photographic archives or by Liz Thach

Chapter Eleven: Diamond Creek Vineyards

Brounstein, A. (2000). "Diamond Creek Vineyards: The Significance of Terroir in the Vineyard," an oral history conducted in 1998 by Carole Hicke, Regional Oral History Office, The Bancroft Library, University of California, Berkeley, 2000.

Legget, H. B. (1941). Early History of Wine Production in California. Wine Institute Report.

NVV. (2012). Napa Valley Appellations. Available at: http://www.napavintners.com/napa_valley/appellations.asp.

Swinchatt, J. & Howell, D.G. (2004). The Winemaker's Dance: Exploring Terroir in the Napa Valley. Berkeley: University of California Press.

The Wine Cellar Insider. Available at: http://www.thewinecellarinsider.com/california-wine/diamond-creek-california-wine-cabernet-sauvignon/

UC Cooperative Extension (2013). 2012 Sample Costs to Establish a Vineyard and Produce Wine Grapes – Napa Valley. Available at: http://www.napagrowers.org/wp-content/uploads/Napa-County-Cost-Study-2012.pdf

Photo Credits: Courtesy of Diamond Creek Vineyard photographic archives or by Liz Thach

Chapter Twelve: Beaulieu Vineyards

BV. (2013). Beaulieu Vineyard and Georges de Latour. Available at: http://www.bvwines.com.

Byington, L. F. (1931). "Biography of Georges de Latour" in History of San Francisco. Chicago: S. J. Clarke Publishing Co. Vol. 3 Pages 344-348.

Chappellet, M. (1998). Gardens of the Wine Country. NY: Chronicle Books

Life (1945). "Life Goes to a Debut". Life Magazine. November 19, 1945.

Prial, F. J. (1990). Wine Talk. The New York Times Archives. Oct 24, 1990. Available at: http://www.nytimes.com/1990/10/24/garden/wine-talk-422390.html

Rayapati, N. A. (2012). Major Grapevine Diseases: Fanleaf and Leafroll. Available at: http://cru.cahe.wsu.edu/CEPublications/FS074E/FS074E.pdf

Rutherford Dust Society (2013). Facts about the Rutherford Dust AVA. Available at: http://www.rutherforddust.org/

USDA (2012). Grapevine Red Blotch Disease. Available at: http://www.ucanr.edu/sites/viticulture-fresno/files/157139.pdf

Photo Credits: Courtesy of BV Vineyard photographic archives or by Liz Thach

Chapter Thirteen: Lessons

Bonne, J. (2014). The New California Wine: A Guide to the Producers and Wines Behind A Revolution in Taste. Berkeley: Ten Speed Press.

CSWA. (2014). California Sustainable Winegrowing Program. Available at: http://www.sustainablewinegrowing.org/

Jones, I. (1942). The Vineyard. Berkeley: Univ. of CA Press.

London, J. (1913). John Barleycorn. Oxford University Press.

Lynch, K. (1988). Adventures in the Wine Route: A Wine Buyer's Tour of France. NY: The Noonday Press.

Zap (2013). The Zinfandel Heritage Project. Zinfandel.org. Available at: http://zinfandel.org/default.asp?n1=18&n2=790&member=

Index

About the Author

Dr. Elizabeth "Liz" Thach, MW is a Professor of Management and Wine Business at Sonoma State University in Rohnert Park, California. Her areas of expertise include Wine Business Strategy, Human Resource Management, and Leadership Development. She started teaching as a full-time professor at SSU in the Fall of 200. In addition to teaching, she currently conducts research and does consulting with multiple wineries and other local businesses. In May of 2011 she was awarded the title of Master of Wine from the Institute of Masters of Wine in London, making her the first female MW on the West Coast and the 7th woman in the US to achieve this honor.

Liz holds a doctorate in Human Resource Development from Texas A&M University (1994); an M.A. in Organizational Communication and Management from Texas Tech University (1987), and a B.A. in English from College of Notre Dame in Belmont, CA (1983). An award winning writer, she has published more than 100 articles; 6 wine books and 8 book chapters. She is an active member in community organizations, and has held positions with the Board of Directors for Women for Wine Sense, Petaluma Gap WineGrowers Alliance, PASCO and many others. She is on the Editorial boards for *International Journal of Wine Business Research* and *Wine Economics & Policy*. In addition, Liz serves as a professional wine judge at many prestigious wine judging competitions.

Liz is married with one daughter. She lives on Sonoma Mountain where she spends her free time hiking, reading, bird-watching, and tending her hobby vineyard from which she makes pinot noir and sauvignon blanc wine (photo above). She and her husband also enjoy golfing, travel, and cooking together, as well as having friends over for wine-tastings and barbeques.

Made in the USA
San Bernardino, CA
20 December 2014